ORIGINAL RESOPAL

JOVIS

ORIGINAL RESOPAL

JOVIS

INHALT
CONTENT

6   Vorwort
Foreword
Peter Cachola Schmal/Romana Schneider

8   Einführung
Introduction
Ingeborg Flagge

10   Resopal – Weit mehr als Laminat
Resopal – Much more than a Laminate
Günter Lattermann

20   KUNSTHARZPRODUKTE AUS PHENOPLAST,
DEM VORGÄNGERMATERIAL VON RESOPAL
PRODUCTS MADE OF PHENOPLAST,
THE SYNTHETIC RESIN FORERUNNER OF RESOPAL

30   RESOPAL ALS PRESSMASSE UND SCHICHTSTOFFPLATTE
RESOPAL AS A MOULDING COMPOUND AND LAMINATE

38   »Vertreibung aus dem ersten Paradies« – Kunststoff in der
Bundesrepublik der fünfziger Jahre im Kontext von Design-
diskursen
"Expulsion from the Primal Paradise" – Plastics in West
German Design Discourses in the 1950s
Gerda Breuer

52   DIE ERSCHAFFUNG DER MARKE RESOPAL DURCH JUPP ERNST
THE CREATION OF THE RESOPAL BRAND BY JUPP ERNST

64   RESOPAL IN ÖFFENTLICHEN RÄUMEN
RESOPAL IN PUBLIC SPACES

92 Leben und Wohnen mit Resopal nach dem
Zweiten Weltkrieg
Living with Resopal after the Second World War
Eva Brachert

102 RESOPAL IM WOHNHAUS
RESOPAL IN THE HOME

132 RESOPAL UND DIE KUNST
RESOPAL AND ART

144 Vorgehängt und hinterlüftet. Oberfläche und Architektur
Clad and Rear-ventilated. Surface and Architecture
Ulrich Höhns

156 RESOPAL UND ARCHITEKTUR
RESOPAL AND ARCHITECTURE

168 Resopal – Das Oberflächenphänomen
Resopal – The Surface Phenomenon
Gert Selle

180 Die Macher der Oberfläche
The Creators of the Surface
Gerd Ohlhauser

192 Anhang
Appendix

193 Anmerkungen
Notes

196 Verwendete Quellen bei den Kurztexten
Bibliography for the Catalogue Sections

196 Abbildungsnachweis
Photo Credits

197 Autoren
Authors

199 Register
Index

200 Impressum
Imprint

VORWORT UND EINFÜHRUNG
FOREWORD AND INTRODUCTION

Die Entstehung des Werkstoffs Resopal ist eng mit der Entwicklung der Moderne in Deutschland verbunden. 1930 patentiert, wurde Resopal bald eins mit dem Phänomen der Kunststoffoberfläche. Es war ein Material, das universal, das heißt in nahezu allen Bereichen des modernen Lebens, einsetzbar war – in der Industrie, im öffentlichen Raum, in den eigenen vier Wänden; und wie bereits vor dem Krieg war es bis in die siebziger Jahre hinein auch eng an neuartige Konsumgüter gebunden. Die Blütezeit von Resopal fällt allerdings in die Zeit des Wiederaufbaus nach dem Zweiten Weltkrieg. Nun schien auch in Deutschland Realität geworden, was in den USA schon in den zwanziger Jahren formuliert wurde, nämlich dass Kunststoff als Vermittler von angewandter Demokratie wirke, indem er Luxusartikel zum Allgemeingut der Massen werden ließ. Die Marke Resopal hat einen dementsprechend großen Anteil am Mythos der Nierentischära. Wie sich der Werkstoff danach und bis heute neuen gesellschaftlichen Entwicklungen angepasst und deren Erscheinungsformen mitgeprägt hat, ist ebenfalls Gegenstand dieses Buches, das die gleichnamige Ausstellung im Deutschen Architekturmuseum (DAM) begleitet.

An dem Ausstellungs- und Buchprojekt, das sich zunächst äußerst schwierig gestaltete, da kaum historische Archivalien vorlagen, waren verschiedene Personen beteiligt, denen für das Gelingen besonderer Dank gebührt. An erster Stelle steht – vertreten durch den Geschäftsführer Donald Schaefer – die Resopal GmbH, die maßgeblich dazu beigetragen hat, dass dieses Projekt realisiert werden konnte. Gerd Ohlhauser, Designchef der Resopal GmbH und Initiator dieses Projekts, hat es mit nie versiegender Energie tatkräftig forciert und hilfreich begleitet. Durch sein Engagement konnten viele Teile eines bis dato unbekannten Puzzles über die Geschichte eines Werkstoffs zusammengetragen werden, der zwar als Synonym der Kunststoffplatte gilt, über den aber sonst wenig Wissen verbreitet ist. Auch Mitarbeitern der Resopal GmbH ist es zu danken, dass noch Zeugnisse zur Geschichte dieses Werkstoffs gefunden werden konnten. Danken möchten wir auch Irene Neudecker, die im Rahmen ihrer an der FH Coburg erstellten Diplomarbeit »75 Jahre Resopal – Ein Material schreibt Kulturgeschichte« in der Anfangsphase des Projekts in Kooperation mit der Kuratorin Rechercheergebnisse und Materialien eingebracht hat. Den Autoren Eva Brachert, Gerda Breuer, Ulrich Höhns, Günter Lattermann und Gert Selle sind wir dankbar, dass sie sich des bisher unzureichend bearbeiteten Themas angenommen haben und hoffen, dass dies zu weiteren Entdeckungen in der Geschichte der Kunststoffe führen wird. Besonders freut uns, dass auch Kunstwerke aus Resopal in der Ausstellung vertreten und im Buch abgebildet sind. Unser Dank gilt den Künstlern, die unser Projekt unterstützt haben, aber auch allen anderen Leihgebern sowie den beteiligten Architekten und Fotografen. Mario Lorenz und seinem Team danken wir für die kongeniale Umsetzung der Inhalte in der Ausstellungsinszenierung und dem Künstler Dieter Balzer für die Erstellung einer speziell für diese Ausstellung entworfenen Installation.

Dass die Qualität historischer Abbildungen in Buch und Ausstellung trotz mancher technischer Hürden dennoch hoch ausfallen konnte, verdanken wir der Firma Lasertype, die das Projekt auch finanziell unterstützt hat. Ebenso haben sich verschiedene, für den Herstellungsprozess von Resopal wichtige Firmen mit eigens für diese Publikation entworfenen Beiträgen für das Projekt engagiert und so den Katalog ermöglicht.

Peter Cachola Schmal
Direktor des Deutschen Architekturmuseums
Romana Schneider
Kuratorin der Ausstellung

The genesis of the material Resopal is very closely linked with the development of Modernism in Germany. Patented in 1930, Resopal soon became indistinguishable from the phenomenon of the plastic surface finish. It was a material with universal application possibilities, i.e. it could be used in virtually all spheres of modern life – in industry, the public realm, or in the home; and from even before the War right through into the seventies it was also closely connected with new kinds of consumer goods. But Resopal enjoyed its heyday in the period of reconstruction following the Second World War. The idea that had been formulated in the USA back in the 1920s now seemed to have become reality in Germany too, namely that plastic acted as a vehicle for democracy in action by making luxury articles the general property of the masses. The Resopal brand thus played a significant role in the myth of the "*Nierentisch*" age". How the material has since then adapted to new developments in society and their manifestations is also the subject of this book, published to accompany the exhibition of the same name at the German Museum of Architecture (DAM).

A number of different people were involved in the exhibition and book project which was initially extremely difficult since there was very little archive material on the subject. They deserve to take special credit for the success of the project. First and foremost, we must mention the company Resopal GmbH – represented by its chief executive Donald Schaefer – whose contribution was crucial to the project's realisation. Gerd Ohlhauser, head of design at Resopal GmbH and initiator of this project, provided helpful back-up and vigorously drove the project forward with his never-flagging energy. Thanks to his commitment, it was possible to bring together many pieces of what had been an unidentified jigsaw of the history of the material, about which there is little knowledge even though it has become a synonym for laminates per se. Our thanks also go to a number of employees of Resopal GmbH for having found records documenting the history of the material. We would further like to thank Irene Neudecker, who in the early stages of the project, as part of her degree dissertation at Coburg University of Applied Sciences on "75 years of Resopal – a material makes cultural history" provided research findings and material in collaboration with the exhibition's curator. We are also grateful to the authors Eva Brachert, Gerda Breuer, Ulrich Höhns, Günter Lattermann and Gert Selle for their willingness to work on a topic that to date has not been adequately explored. We hope that this will lead to further discoveries in the history of plastics. We are delighted that the exhibition includes a number of works of art in Resopal, photographs of which are included in the catalogue. Our thanks go to the artists who supported our project, and also to everyone else who loaned items for the exhibition and to the architects and photographers involved. We thank Mario Lorenz and his team for their inspired transformation of abstract ideas into a tangible exhibition and the artist Dieter Balzer for making an installation specially for this exhibition.

Lasertype must take credit for the fact that the quality of the historical illustrations in both the book and the exhibition have remained of the highest calibre, despite a number of technical hurdles. The company also supported the project financially. A number of companies that are important for the process of manufacturing Resopal also lent their support, making contributions that were specially designed for this book and which made the catalogue possible.

Peter Cachola Schmal
Director of the German Museum of Architecture
Romana Schneider
Curator of the exhibition

Es bedarf eines gewissen Alters, um sich an die farbfreudigen Resopal-wohnwelten der fünfziger und sechziger Jahre zu erinnern, an rosa Blumenhocker, an *bleue* Nierentische, an grau-weiß gesprenkelte Schrankwände. »Resopal ist ideal« hieß es damals. Und auch wenn nicht jeder das leicht-fröhliche Material liebte, es wertete die Küche zum »Reich der Frau« auf (wie Eva Brachert in ihrem Aufsatz darlegt) und genoss durchaus einen Statuscharakter in Deutschland, als am Ende des Zweiten Weltkrieges die meisten alles verloren hatten und sich nun wieder für eine Zukunft einrichteten, die »Resopal gestaltet«.

Neben dem gewissen Charme, den das Material hatte, war es auch sehr praktisch: leicht zu reinigen, widerstandsfähig gegen Chemie, auch mit Gewalt kaum kaputt zu kriegen. Kurz, es war der Stolz jeder Hausfrau, die Altern schäbig findet und Gebrauchsspuren hasst.

Resopal, das bedeutete zu seiner Hoch-Zeit eine Gattungsbezeichnung wie Uhu und Tempo. Aber wusste wirklich jeder, wie Günter Lattermann schreibt, was genau dahinter steckte? Nämlich, wie es in Meyers Lexikon heißt, ein Handelsname für Kunststoffplatten, »die aus einer Trägerschicht und einer Deckschicht bestehen, die bei hohem Druck zusammengepresst werden« und dabei »leicht abwaschbar, unempfindlich gegen Säuren und Laugen sowie widerstandsfähig gegenüber mechanischer Beanspruchung« sind.

Genügen solche Eigenschaften, um Resopal eine Ausstellung im Deutschen Architekturmuseum zu widmen? Natürlich nicht, aber dass ich als Direktorin spontan zusagte, als Gerd Ohlhauser das Thema vorschlug, hatte andere Gründe.

Erstens liest sich die Geschichte der Erfindung und Herstellung von Resopal wie ein sehr deutscher historischer Krimi. Darüber hinaus steht Resopal für einen ideologischen Streit, dem Gerda Breuer in diesem Buch nachgeht. Während Architekten im Westen, wie Hans Schwippert, von einer Charakterlosigkeit des Materials sprachen und es als Surrogat und Imitat denunzierten, erlebte es als ein wichtiges Material für Artikel aus Plaste in der DDR Hochkonjunktur.

Mir schien es wichtig, die »bewegte, ja durchaus überraschende und überzeugende Geschichte« (Gert Selle) des Materials der interessierten Öffentlichkeit einmal genauer vorzustellen. Und zwar eben aus den Gründen, die ein Museum für Kunst und Design dazu verleiteten, das Angebot einer Einrichtung der Ausstellung abzulehnen, weil »es sich eindeutig um ein kultur- und industriegeschichtliches Ausstellungsthema handelt«. (...) »Es mag Ihnen der Unterschied zwischen den beiden Aspekten nicht so gewichtig erscheinen wie uns. Aber Kunstmuseen sehen ihre Aufgabe darin, Kunst- und Formgestaltung als eine vorbildliche historische Leistung darzustellen, während kultur- und industriegeschichtliche Museen zeittypische Situationen und technische Innovationen vermitteln. Mit Kunst und Design hat das Material Resopal nur in ausgewählten Einzelbeispielen zu tun.«

Ein weiterer Grund für die Entscheidung, dass der Werkstoff für die Besucher des DAM interessant sein könnte, liegt in der Entwicklung der Architektur selbst. Die Materialität eines Hauses – vieldeutiger Begriff, wie Ulrich Höhns schreibt – ist zunehmend das Thema des architektonischen Entwurfs. Bei der unglaublichen Vielfalt an Oberflächen, die es heute gibt und die fast alles können, ist die Wahl des Materials eine Sache der Wirkung, die ein Architekt erzielen will. Nachdem es die »archaische Wand, die schützt (und) trägt« (Höhns) nicht mehr gibt und die sogenannte Ehrlichkeit in der Architektur kaum noch eine Rolle spielt, geht es um Effekte. Wo die Konstruktion nicht spektakulär ist oder sein darf, kann es immerhin die Oberfläche sein.

Resopal kann heute alles; es kann jeden anderen Werkstoff simulieren. Dies geht so weit, dass das hochwertige und heute teure Material im Ver-

You need to be of a certain age to remember the brightly coloured Resopal domestic interiors of the 1950s and 60s, the pink flower stools, the kidney-shaped coffee tables, the grey-and-white speckled wall units. "Resopal is ideal" was the slogan of the time. And although not everyone was enamoured with this light, cheerful material, it nevertheless upgraded the kitchen to the "woman's kingdom" (as described by Eva Brachert) and enjoyed a certain status in Germany at a time when most people had lost everything during the Second World War and were now trying to set up home again using Resopal as a major design feature.

But the material itself not only had a certain appeal, it was also practical: easy to clean, resistant to chemicals and well nigh impossible to wear out. In short: it was the pride and joy of every housewife who equated age with shabby and hated things to bear signs of wear.

In its heyday, Resopal was a generic name like Sellotape or Hoover. But, as Günter Lattermann writes, did everyone really know exactly what it was? – Namely, as Meyers Lexikon states, a trade name for laminates "consisting of a substrate and a top layer that are fused together under high pressure. Resopal laminates are easy to wipe down, resistant to acid or alkaline substances and to mechanical wear and tear."

However, are these properties enough to warrant dedicating an entire exhibition in DAM (German Museum of Architecture) to Resopal? Of course not: there are quite different reasons to explain why as director I spontaneously agreed to Gerd Ohlhauser's suggestion to stage this exhibition.

First, the story of Resopal's invention and manufacture reads like a historical novel – and a very German one at that. Furthermore, Resopal represents an ideological dispute, which Gerda Breuer traces in this book. Whereas in West Germany architects such as Hans Schwippert spoke of the material's lack of character and denounced it as a surrogate and imitation, it enjoyed booming popularity in East Germany as an important Plaste article.

I felt it was important to analyse the "eventful, thoroughly surprising and convincing history" (Gert Selle) of this material. My reasons were in fact the very same reasons that led a museum of art and design to turn down the offer to stage an exhibition that "is clearly (...) concerned with cultural and industrial history." (...) "The difference between the two aspects may not seem as important to you as to us. But art museums believe that their role is to depict art and design as an exemplary historical achievement, whereas museums of cultural and industrial history convey situations and technical innovations that are typical of their time. The material Resopal is related to art and design in only a small number of individual cases."

Another important reason for deciding that this material might be interesting for visitors to DAM is connected with the development of architecture itself. The materiality of a building – a term that, as Ulrich Höhns writes, is open to interpretation – has increasingly become a topic of interest to architectural design. Given the incredible variety of surface finishes available today that are capable of virtually anything, the choice of material is a matter of the effect an architect wishes to create. Now that the "archaic wall that carries loads" (Höhns) no longer exists and so-called honesty in architecture scarcely plays a role anymore, the main concern is effects. The surface of a building can be spectacular even when the structure itself is not, or cannot be.

Nowadays Resopal can do anything and everything; it can simulate any other material. Indeed, this expensive high-calibre material that no longer looks like a cheap downmarket product, transposes everything that is physical and tangible onto the façade – as "a fake on the surface, where it becomes the image of something that is not actually there" (Selle).

gleich zu dem ehemaligen Billigprodukt an der Fassade das Körperlich-Greifbare der Architektur als »Fake auf die Oberfläche, die zum Bild von etwas wird, das gar nicht da ist« (Selle) verlagert. Ulrich Höhns spricht dabei von einer Entgrenzung des Raums, Gert Selle von einem »Verwirrspiel mit der Wahrnehmung«, das die postmoderne Entwicklung des Materials heute erlaubt.

Nachdem Resopal in den letzten Jahrzehnten eine »Durststrecke der Reizarmut« (Selle) hinter sich gebracht hat, steht es gegenwärtig für fast jede Lösung in der Architektur zur Verfügung, von einer Oberfläche, die transluzent wirken möchte bis hin zu tapetenhaftem Oberflächendesign, das der Fassade eines Baus die Anmutung eines nach außen gekehrten Innenraums gibt. Wenn das kein Thema für das DAM ist.

Trotz solcher Vielfältigkeit gibt es nur wenige gute Architekturlösungen mit Resopal. Vielleicht erkennt man sie auf Fotos ohne Materialangaben auch einfach nur nicht, weil Resopal so viele andere Materialien assoziiert und zitiert. Jedenfalls ist Resopal ein breites und faszinierendes Thema, und ich hoffe, dass die Besucher der Ausstellung und die Leser dieses Buchs dies nachvollziehen werden.

Ingeborg Flagge
Ehemalige Direktorin des Deutschen Architekturmuseums

Ulrich Höhns talks of it "abolishing the defining edges of spaces," Gert Selle of "the play with perception," which the postmodern evolution of Resopal now permits.

Having gone through "a long arid stretch when it was pretty much devoid of appeal" (Selle) Resopal is now available to solve almost any problem in architecture, be it a covering that would like to be translucent or a wallpaper-like surface design that makes a building's external space look like an interior space turned inside out. Surely that is a topic of interest to DAM?

Despite this great range of possibilities available, there are not as many good architectural solutions that use Resopal as you might imagine. Perhaps we just do not recognise them on photos that are published without details of the materials used because Resopal mimics so many other materials, pretending to be something it is not. At any rate, Resopal is a fascinating topic that covers a broad spectrum of interest and I hope that visitors to the exhibition and readers of this book will be able to get a sense of that.

Ingeborg Flagge
Former director of the German Museum of Architecture

Günter Lattermann

RESOPAL – WEIT MEHR ALS LAMINAT
RESOPAL – MUCH MORE THAN A LAMINATE

Resopal: Heute weiß jeder sofort was damit gemeint ist. Der Name ist so bekannt, dass er als Gattungsbezeichnung für eine komplette Produktgruppe in die Alltagssprache übernommen worden ist. So wie Tempo für Taschentücher oder Fön für Haartrockner, UHU für Klebstoffe, Tesa für Klebeband oder Frigidaire in Frankreich für Kühlschränke. Sagt man Resopal, meint man Schichtstoffplatten. Solche Platten bestehen aus mehreren Papierbahnen, die mit Kunstharz imprägniert und unter Hitze und hohem Druck verpresst sind. Techniker sprechen auch von Laminaten (von lat.: lamina, die Schicht). Im angelsächsischen Bereich beobachten wir das gleiche Phänomen unter einem anderen Namen: Dort bezeichnet man Kunstharz-Schichtstoffe oft allgemein als Formica.

Woher kommt nun der Begriff Resopal, wer hat ihn geprägt und wo liegt sein Ursprung? Der Produktname Resopal ist eingeführt worden von einer der wichtigsten Kunstharzfirmen der Vorkriegszeit, der H. Römmler AG in Spremberg in der Niederlausitz. Der Begriff Resopal stand am Anfang allerdings keineswegs für das Produkt Schichtstoffplatte, sondern für eine bestimmte, neu entwickelte Klasse von Kunststoffen, die Aminoplaste oder Harnstoffharze. Diese neuartigen Materialien hatten für die damalige Zeit so wunderbare Eigenschaften wie brillante Farbigkeit, Widerstandsfähigkeit gegen Chemikalien und Temperatur, Lebensmittelechtheit und große Härte, dass sie zunächst für viele Haushaltsgegenstände des täglichen Bedarfs, Teller, Tassen, Schüsseln, Schalen etc. verpresst wurden. Dabei übernahm ein »Formgestalter« (heute würde man sagen Designer) die Formgebung der Produkte. Die zwei Aspekte, nämlich dass Resopal ursprünglich weit mehr als Laminat war und dass die Anfänge des Produktdesigns von Kunststoffen in Deutschland in die Zeit der beginnenden dreißiger Jahre fielen, gerieten aus verschiedenen Gründen nach dem Zweiten Weltkrieg fast völlig in Vergessenheit. Warum? Dazu müssen wir etwas näher in die Firmengeschichte der damaligen H. Römmler AG hineinschauen.

Im Jahre 1867 gründete August Hermann Römmler in Spremberg eine Fabrik, die H. Römmler GmbH, in der Reststoffe der Tuch- und Hutfabrikation zu Scherhaarmaterialien für die Textilindustrie aufbereitet wurden.[1] 1904 wurde die Produktpalette erweitert und Baumwollflocks als Füllstoffe für die junge Schallplattenindustrie hergestellt. Als Grundstoff für die neuen Tonträger war, nach ersten Versuchen mit Celluloid und dem hochvulkanisierten Hartgummi Vulcanit, ab 1896 eine in Wärme pressbare Mischung aus hauptsächlich Schellack und eben den Baumwollflocks als Füllstoff verwendet worden.[2] Alle drei Schallplattenmaterialien sind Vorläufer synthetischer Kunststoffe. Gefüllte Vulcanit- und Schellack-Mischungen wurden aber auch in großem Ausmaß als Isolierstoffe für die ebenfalls junge Elektroindustrie gebraucht, ein Gebiet, in das Hermann Römmler jun. 1905 erfolgreich einstieg. Der ursprüngliche Betrieb wurde unter den Brüdern aufgeteilt. Arthur und Kurt Römmler führten die Scherhaarabteilung als WeKa-Tuchpapierfabrik A. & K. Römmler AG fort. Hermann Römmler jun. entwickelte die elektrotechnische Abteilung als H. Römmler GmbH zu einer der führenden Fabriken für plastische Massen,[1,3] nachdem er sich parallel zu Leo Hendrik Baekeland mit der Entwicklung neuer vollsynthetischer Kunstharze, den Phenol-Formaldehyd-Harzen beschäftigt hatte. 1907 meldet Baekeland sein Phenoplast Bakelit zum Patent in den USA an, 1908 patentierte Römmler sein Phenoplast Hares (aus *Hermann Römmler Spremberg* zusammengesetzt). Nach längeren Patentstreitigkeiten kam es 1919 zum Vergleich: Die seit 1917 als H. Römmler AG firmierende Gesellschaft in Spremberg besaß danach ein lizenzfreies Mitbenutzungsrecht der Patente von Bakelit,[4] das ab 1910 als erster synthetischer Kunststoff in Erkner bei Berlin produziert wurde. Die H. Römmler AG gehörte bis zum Zweiten Weltkrieg zu den acht großen Unternehmen in Deutschland,

Resopal: today everyone immediately realizes what that is. The name is so well known that it has become a generic name for an entire group of products in common parlance – just as in British English Kleenex is used for tissues, Hoover for vacuum cleaner or Sellotape for sticky tape, or *Frigidaire* is the common word in France for refrigerator. When someone says Resopal, he means laminate. These laminates consist of several layers of paper that have been impregnated with synthetic resin and pressed together under high pressure. The word laminates comes from the Latin *lamina* meaning a layer. In the Anglo-Saxon world the same phenomenon can be seen with a different name: laminates impregnated with synthetic resin are often generally referred to as Formica.

Where does the name Resopal come from, who coined it and what is its etymology? The product name Resopal was introduced by one of the most important companies making synthetic resin before the Second World War, H. Römmler AG in Spremberg in the Niederlausitz region. However, Resopal did not initially stand for the laminated panels, but for a particular, newly developed group of plastics – aminoplasts or urea resins. These new materials had such wonderful properties – they came in vivid colours, were resistant to chemicals and temperature, were food safe, and very hard – that they were initially moulded to make predominantly everyday household articles – plates, dishes, bowls, cups etc. A designer was commissioned for these plastic articles. The two aspects, namely that Resopal was originally far more than a laminate and that product design was applied to synthetic materials in Germany for the first time in the early 1930s – were almost forgotten after the Second World War for different reasons. Why? To answer that question, we will have to take a closer look at the history of the company H. Römmler AG at that time.

In 1867, August Hermann Römmler founded a factory in Spremberg, H. Römmler GmbH, in which he processed leftover material from the cloth and hat manufacturing industries into shear fibres for the textile industry.[1] In 1904 the company extended its product range and produced cotton fibres as filler material for the gramophone record industry, which was still in its early days. Following initial experiments with Celluloid and the highly vulcanised hard rubber Vulcanite, the basic material for records from 1896 onwards was a hot-moulded mixture made mainly from shellac with cotton fibres as a filler.[2] These three materials for gramophone records are the precursors of synthetic plastics. But filled Vulcanite and shellac mixtures were also needed in greater quantities as insulating materials for the electrical industry, which was in its early days, a field which Hermann Römmler jun. successfully entered in 1905. The original company was divided up amongst the brothers. Arthur and Kurt Römmler continued the "Scherhaarabteilung" (shear fibres department) as "WeKa-Tuchpapierfabrik A.&K. Römmler AG," Hermann Römmler jun. developed the electro technical department into one of the leading manufacturers of plastic compounds,[1,3] having turned his attention to the development of new completely synthetic resins – the phenolic formaldehyde resins – at the same time as Leo Hendrik Baekeland. In 1907 Baekeland filed for a patent for his phenoplast Bakelite in the USA; in 1908 Römmler filed an application for a patent for his phenoplast Hares (from *Hermann Römmler Spremberg*). After protracted disputes a settlement was arbitrated in 1919: the Spremberg-based company that had gone under the name of H. Römmler AG from 1917 obtained joint use rights on the Bakelite patents[4] which from 1910 onwards was produced in Erkner near Berlin as the first synthetic plastic. Until the Second World War, H. Römmler AG was one of Germany's eight large companies producing phenoplasts. In Spremberg they were manufactured in the Römmler

Ab 1905 Zusammenarbeit mit Beka (Bumb & König), Berlin
From 1905 in collaboration with Beka (Bumb & König), Berlin

Hermann Römmler jun.,1881–1949
Hermann Römmler jun., 1881–1949

die Phenoplaste herstellten.[5] Die Spemberger Harze stammten aus den Römmlerschen »Chemischen Fabriken«. Ihre Haupteinsatzgebiete waren zum einen Formteile aller Art, die im »Presswerk« gefertigt wurden, zum anderen erzeugte die »Hartpapierfabrik« geschichtete Pressstoffe (Laminate) und die »Lagerschalen-Abteilung« Gleitlagerschalen und -büchsen.[6,7] Für die fast ausschließlich im technischen Bereich eingesetzten Phenoplast-Schichtpressstoffe für Hartpapier-Laminate und vor allem auch für Gleitlagerschalen besaß die H. Römmler AG eine von Anfang an (d. h. ab 1920[8]) führende Herstellerposition in Deutschland.[9] Was Formpressteile aus Phenoplast betrifft, die auch zum Fabrikationsprogramm von zahlreichen anderen Kunstharz-Presswerken gehörten, so wurden in Spremberg elektrotechnische Artikel wie Radiogehäuse (zum Beispiel der Volksempfänger),[10] Telefongehäuse,[11] und die erste vollständig aus einem Kunststoff bestehende Schreibtischleuchte[12] hergestellt. Diese war um 1929 von Christian Dell, dem ehemaligen Werkmeister der Bauhaus-Metallwerkstatt und einem der später bekanntesten und erfolgreichsten Pioniere der Leuchtengestaltung, entworfen worden. In diese Zeit fällt auch die Geburtsstunde von Resopal.

Man war in der Kunststoffindustrie schon seit einiger Zeit mit der Entwicklung neuer Harze beschäftigt, die im Gegensatz zu den dunklen, braunen bis roten Phenoplasten zunächst farblos und daher völlig variabel einzufärben waren, mit entsprechenden Pigmenten sogar reinweiß hergestellt werden konnten. Dies war bislang nicht möglich gewesen. Erstmalig wurden solche Kunstharztypen 1925/26 in Großbritannien industriell produziert.[13]

Ähnliche Typen wurden in Deutschland unabhängig voneinander 1927/28 von der Dynamit AG Troisdorf[14] und 1929/30 von der H. Römmler AG[15] hergestellt. Letztere stellte erste Tischgeräte aus hellfarbigem Aminoplast 1930 auf der Leipziger Frühjahrsmesse aus.[16] Allerdings hatte der neue, sensationelle Pressstoff noch nicht den später so berühmten Namen Resopal. Zunächst hieß er Alboresin.[17,18] Dies bedeutet »weißes Harz« (lat.: *resina alba*). Erst auf der Leipziger Frühjahrsmesse 1931 tauchte zum ersten Male der Name Resopal auf.[19] Es ist bislang nicht bekannt, weshalb der Name gewechselt wurde oder werden musste. Es kann lediglich vermutet werden, dass womöglich eine der acht konkur-

"Chemicals Factories."[5,6] The main application for phenolic resins and phenoplasts was mouldings of all kinds, which were manufactured in the "Moulding Press Department". On the other hand, the "Laminated Paper Mill" produced laminated products (laminated papers or fabric) and the "Bearing Bushing Department" manufactured sliding bearings and bushings.[7] For the phenoplast laminates, the laminated papers and especially for sliding bearings that were used almost exclusively for engineering applications, H. Römmler AG occupied the position of leading manufacturer[8] in Germany from the outset (i.e. from 1920[9]). With regard to phenoplast mouldings, which were also part of the range of numerous other synthetic resin moulded products, electrical articles such as radio housings (e.g. for the *Volksempfänger*),[10] telephone housings[11] and the first desk lamp to be made entirely from a synthetic material[12] were manufactured in Spremberg. The latter had been designed in 1929 by Christian Dell, former metal workshop master at the Bauhaus and later one of the most famous and successful pioneers of lamp design. The birth of Resopal also occurred during this time.

The plastics industry had been working for some time on developing new resins, which unlike the dark phenoplasts which ranged from red through to brown in colour, were initially colourless and could therefore be dyed any colour or using the right pigments could even be pure white. This had not previously been possible. These types of synthetic resin were manufactured industrially for the first time in 1925/26 in Great Britain.[13]

Similar types were produced independently of each other in Germany in 1927/28 by Dynamit AG Troisdorf[14] and in 1929/30 by H. Römmler AG.[15] This company exhibited the first table-top appliances in pale coloured aminoplast at the 1930 Leipzig Spring Fair.[16] But this sensational new moulded material did not yet have the name Resopal that would become so famous. Initially it was called *Alboresin*[17,18] meaning "white resin" (Latin: *resina alba*). It was not until the Leipzig Spring Fair in 1931 that the name Resopal appeared for the first time.[19] No-one knows why the name was changed, or perhaps had to be changed. We can only surmise that one of the eight competitors making laminates in Germany perhaps had protested vehemently. In Germany it was (and still is) com-

Walter Maria Kersting: Volksempfänger VE 301
aus dem Phenoplast Hares, Entwurf 1933
*Walter Maria Kersting: Volksempfänger VE
301 made of the phenoplast Hares, designed
in 1933*

Telefon der Reichspost aus dem Phenoplast Hares
Reichspost telephone made of the phenoplast Hares

Christian Dell: Schreibtischleuchte aus dem
Phenoplast Hares, Entwurf 1929
*Christian Dell: desk lamp made of the phe-
noplast Hares, designed in 1929*

rierenden Pressstoffhersteller in Deutschland Einspruch erhob. Es war
(und ist noch) in Deutschland verbreitet, in die Namen von chemischen
Produkten, insbesondere von Kunststoffen, die Firmen- oder Ortsnamen
einzubinden. So hat die damalige Firma Albert die Phenoplast-Lackharze Albertol, Albertit, Albamit und Alresat hergestellt, die noch heute –
inzwischen von anderen Firmen – produziert und vertrieben werden.[20,
21, 22] Das Römmlersche Alboresin könnte zu verwandt geklungen haben.
In Spremberg dürfte man demnach gezwungen gewesen sein, schnells-
tens einen Ersatznamen zu finden. So könnte man auf Resopal (von lat.:
resina pallida; »farbloses, blasses, fahles Harz«) gekommen sein. Außerdem stecken in der Silbe »Res« noch die Initialen von *Römmler Spremberg* (siehe oben: Hares; vergleiche auch mit dem Firmenlogo HRS, das
vor und auch noch kurz nach dem Zweiten Weltkrieg benutzt wurde[23]).
Diese Erklärung für die Entstehung des Namens Resopal ist wahrschein-
licher als die Herleitung aus einem »opalisierenden Harz«[24], denn die
Namen Resoplan, Resofloor oder Resoform[25] sind sicher nicht von »Opal«
abgeleitet. Überdies gibt es noch viele andere Produktnamen mit einem
Verbindungs-o, wie zum Beispiel Diolen, Moltopren, Styropor. Auf diese
etwas ungewöhnliche Weise erblickte also der Name Resopal das Licht
der Welt. Und es war ihm nicht in die Wiege gelegt, bedeutungsgleich
mit Schichtstoff zu sein.

Zunächst wurde nämlich Resopal als neuer Formpressstoff bekannt
gemacht. Formpressstoffe sind Werkstoffe für Artikel, die in einer Form
unter dem hohen Druck spezieller Pressen hergestellt werden. Die derart
aus Resopal-Formmassen produzierten »vornehm farbigen« Gebrauchsartikel seien im Gegensatz zu solchen aus stets dunklen Phenoplast-
Pressstoffen »materialecht, unempfindlich gegen Fette, Seife, Hitze und
Wasser, stoßfest und leicht zu reinigen«, wie die ersten Werbeanzeigen
von 1931 verkündeten.[26] Neben den »satten Farben und dem natürlichen Spiegelglanz« sei es auch »die Schönheit der modernen Zweckform«,[27] die das neue Material so attraktiv machten bzw. machen sollten. Das letztere Attribut klingt nicht von ungefähr nach Zielen des
Werkbundes (»funktionale Gegenstände von hoher ästhetischer Qualität«[28]) und Grundsätzen des Bauhauses (Walter Gropius: »Ein Ding (...)
soll seinem Zweck vollendet dienen (...) und schön sein«[29]). Die deut-

mon practice to integrate company or place names into the name of
chemical products, especially synthetics. For example, a company named
Albert produced a number of phenoplast coating resins called Albertol, Albertit, Albamit and Alresat that are still being produced and sold
today by other companies.[20, 21, 22] Römmler's *Alboresin* may have sound-
ed too similar. Maybe the company in Spremberg was forced to quickly find a substitute name for its "white resin." Perhaps that is how they
arrived at the name Resopal (from the Latin: *resina pallida*; "pale, colourless, pallid resin"). The initials of the company *Römmler* and the place
Spremberg are also in the syllable "Res" (cf. above: *Hares*; or the company logo HRS which was used before and for a short time after the Second World War.[23]) This explanation for the origins of the name Resopal
seems more probable than the derivation from "opalescent resin".[24] Furthermore, it is highly unlikely that the names Resoplan, Resofloor and
Resoform[25] have got anything to do with "opal". There are also many other German product names where the two parts of the word are linked
by an "o," – Diolen, Moltopren, Styropor. Thus the name Resopal came
into being in this somewhat unusual way. And it was not destined from
the outset to be synonymous with laminate.

First of all Resopal became known as a new moulding compound. Moulding compounds are materials used for articles that are manufactured in
a mould under high pressure using special presses. The "stylishly colour-
ful" articles of everyday use made in that way from Resopal moulding
compounds were in contrast to the phenoplast moulding compounds,
which were always dark, "true to the material, resistant to grease, soap,
heat and water, impact resistant and easy to clean", as the first advertisements proclaimed in 1931.[26] Apart from the "rich colours and mirror finish"
it was also "the beauty of this modern functional form",[27] which made, or
would make, the new material so attractive. It is no coincidence that the
latter attribute sounds like the objectives of the Werkbund ("functional objects of high aesthetic quality"[28]) and the principles of the Bauhaus
(Walter Gropius: A thing ... [must] serve its purpose perfectly ... and [be]
beautiful"[29]). In the late 1920s/early 1930s, the German plastics industry
had commissioned designers to design articles of daily use and appliances for the first time. There were a number of different reasons for that.

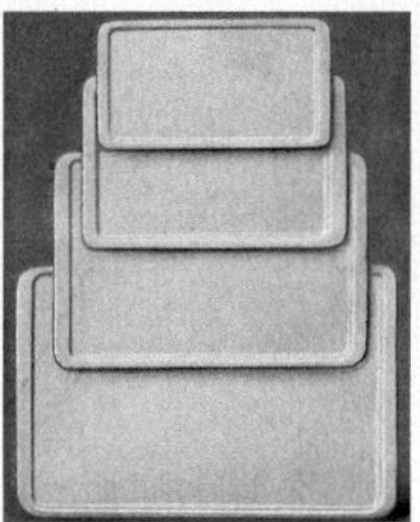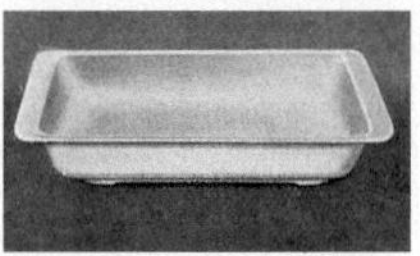

Christian Dell,
1893–1974
Christian Dell,
1893–1974

Service-Einzelteile, Originalfoto Archiv Dell, um 1930
Pieces from a dinner service, original photo, Dell Archive, approx. 1930

Christian Dell: Kaffeetassen und Teller, Originalfoto Archiv Dell, um 1930
Christian Dell: coffee cups and plates, original photo, Dell Archive, approx. 1930

Christian Dell: Vorlegeplatten, Vorlegeschüssel und Vorspeisenplatte aus dem Katalog Resopal Kerit, 1935
Christian Dell: serving platter, serving dish and hors d'œuvres platter from the Resopal Kerit catalogue, 1935

sche Kunststoffindustrie hatte nämlich Ende der zwanziger/Anfang der dreißiger Jahre erstmals Formgestalter mit Entwürfen für Gebrauchsartikel und Geräte beauftragt. Dies geschah aus verschiedenen Gründen. Zunächst herrschte einerseits eine große, lebhafte Aufbruchstimmung in der Kunststoffindustrie. Erstens war die Monopolstellung der Bakelit-Patente hinsichtlich Herstellung und Verarbeitung von Phenoplasten ab 1927 bis 1931 ausgelaufen[30] und damit der Weg für neue Entwicklungen und Aktivitäten freigeworden. Zweitens hatte man gerade die neuen Aminoplaste (Harnstoffharze) entwickelt und konnte mit ihnen ganz neue Produktgruppen erschließen. Andererseits herrschte aber auch tiefe Depression, denn Deutschland geriet im Winter 1929/30 in den Strudel der sich aus dem Zusammenbruch der New Yorker Börse entwickelnden Weltwirtschaftskrise. Ermuntert von der ersten und gezwungen von der zweiten Entwicklung, suchte man in der Kunststoffindustrie nach neuen Wegen der Produktpräsentation und -vermarktung und benutzte die Wirksamkeit der Formgestaltung der Artikel, heute würde man sagen: das Produktdesign als Komponente des Marketing. Dies wurde zudem gefördert durch die neuen Prinzipien der Formgebung, die gerade zu dieser Zeit am Bauhaus oder in ähnlichen Institutionen entwickelt wurden.

Die H. Römmler AG in Spremberg beauftragte Christian Dell – zu dieser Zeit Leiter der Metallwerkstatt an der Kunstschule Frankfurt am Main (Städelschule)[31] – mit der Gestaltung von Pressstoffartikeln. Neben der erwähnten Phenoplast-Leuchte waren dies ab 1929 zahlreiche Haushaltsartikel aus den neuen Aminoplast-Formmassen. 1930 taucht in der *Schaulade*[32] zum ersten Mal eine kleine Schüssel auf, die nach einer Originalfotografie aus dem Archiv Dell eindeutig zuzuordnen ist.[33]

Auf der gleichen Originalfotografie ist ein Tablett zu sehen, das zu der Geschirrserie »Standard« gehört. Sie ist im Katalog »Resopal Kerit« der Verkaufsorganisation Plastica GmbH, Berlin abgebildet.[34] Die H. Römmler AG hatte sich 1932 mit anderen Firmen zu dieser Organisation zusammengeschlossen.[35] Unter der »Standard«-Serie sind im Katalog einige Tabletts bzw. Vorlegeplatten, ein Brotkorb bzw. eine Vorlegeschüssel und eine Vorspeisenplatte aufgeführt, die alle die gleichen Gestaltungsmerkmale aufweisen.

First of all, there was on the one hand a powerful and vibrant mood of new beginnings in industry. Firstly, the monopoly on manufacturing phenoplasts between 1927 and 1931 under the Bakelite patents had run out.[30] This opened up the way for new developments and activities. Secondly, the new aminoplasts (urea resins) had been developed, giving access to completely new product groups. But, on the other hand, the country had sunk into a deep depression: in the winter of 1929/30 Germany had been dragged into the whirlpool of the world economic crisis precipitated by the Wall Street Crash. Encouraged by the first development and obligated by the second, the plastics industry was looking for new ways of presenting and marketing its products. It used the effectiveness of design in selling articles – today we would say the product design as a component of marketing. This was further encouraged by new design principles that were being developed at this time at the Bauhaus and similar institutions.

H. Römmler AG in Spremberg commissioned Christian Dell, – at the time head of the metal workshop at the Art School in Frankfurt am Main (Städelschule)[31] – to design articles of moulded compounds. Apart from the phenoplast desk lamp mentioned above, these included from 1929 onwards, numerous household articles made from the new aminoplast-moulding compound. In 1930 a small dish appeared for the first time in the *Schaulade*[32], which can clearly be identified by an original photograph from the Dell archive.[33]

On the same original photograph a tray can also be seen, which was part of the "Standard" series of tableware. There is a photograph of it in the catalogue "Resopal Kerit" produced by the sales organisation Plastica GmbH, Berlin.[34] In 1932, H. Römmler AG had joined forces with other companies to set up this organisation.[35] The catalogue contains other items from the "Standard" series, including trays, serving platters, a bread basket and serving dishes, and an hors d'oeuvres platter.

Using original photographs from the Dell archive,[36] a dish and coffee service with coffee pot, milk jug, cups and saucers and breakfast plates can also be identified. There was a sugar bowl, butter dish, mustard pot and decorative and powder boxes in the same series. According to the catalogue, all the articles were available in pure white, ivory, dark red, tango (orange-red), yellow, pale green, pale blue, red, yellow and opal.

Anhand von Dellschen Originalfotografien[36] sind weiterhin eine Schüssel und ein Kaffeeservice mit Kaffeekanne, Milchkännchen, Tassen mit Untertassen und Frühstückstellern zu identifizieren . In der gleichen Entwurfsreihe stehen eine Zuckerschale, Butterdosen, Senfgefäße und eine Zier- bzw. Puderdose. Alle Gegenstände konnten laut Katalog in den Farben reinweiß, elfenbein, dunkelrot, tango (orangerot), gelb, hellgrün, hellblau, rot, gelb und opal geliefert werden.

Schließlich ist noch das Dellsche Sportservice zu nennen, das als »große Spitzenleistung« und »Resopal-Wunderkanne« bezeichnet wurde.[37] Sowohl im Hinblick auf die Geschichte der Kunststoffgestaltung bzw. des industriellen Designs als auch von der Entwicklung des Stapelprinzips her nimmt sie eine ganz besondere Stellung in der Designgeschichte ein.[38,39] Weitere Entwürfe von Dell für Haushaltsartikel aus Resopal-Formpressstoff der H. Römmler AG sind sehr wahrscheinlich, aber bislang nicht nachzuweisen.

Nicht nur für die Gestaltung von Haushaltsartikeln, sondern auch für die Werbung engagierte die H. Römmler AG bereits vor dem Kriege professionelle Gestalter. Ab 1930 entwarf Jupp Ernst, damals Grafiker in Bielefeld, für die Firma Römmler AG Werbeprospekte.[40,41]

Die Resopal-Formmasse wurde aber nicht nur für diese frühen Beispiele von Kunststoffgestaltung eingesetzt. Sie war auch wichtig für neue Eigenschaften von Elektroschaltern, Leuchten und Lampenschirmen. In einem Artikel von 1932 über »Die Bedeutung der weißen Pressmasse Resopal für die Gestaltungsmöglichkeit moderner Schaltapparate und Beleuchtungskörper«[42] wird hervorgehoben, dass »die neue Schnellpressmasse Resopal« insofern »einen bedeutsamen Fortschritt« bringe, »als es nun möglich sein wird, einerseits an Stelle der bisherigen auf dunkle Töne beschränkten Schaltapparate« (Schalter) »auch solche in hellen Farben herzustellen, andererseits infolge der Durchlässigkeit der neuen Masse Lampenschirme und Glockenleuchter in diesem Material auszuführen«. Und weiter: »Gerade die ausgesprochen dunkle Tönung muss in ausgesprochen hellen Räumen auffallen. Durch Kombination mehrerer Schaltapparate, wie es heute notwendig ist, entstehen beträchtliche Flächen, die als störend empfunden werden, sobald sie auf hellem Untergrund erscheinen«. »Besonders in Fällen, wo sich etwa aus technischen

Finally, Dell's stacking service, the "Resopal *Wunderkanne*" (magic coffeepot) should be mentioned, which was acclaimed as an "outstanding achievement".[37] Both in terms of the history of plastics design and industrial design and with regard to the development of the stacking principle, it occupies a very special place in the history of design.[38,39] It is likely that Dell designed other household articles in Resopal moulding compound made by H. Römmler AG, but to date there is no known proof.

Even before the War, H. Römmler AG engaged professional designers not only for household articles but also for advertising. From 1930 Jupp Ernst, then a graphic artist in Bielefeld, designed advertising brochures for Römmler AG. [40,41]

However, the Resopal moulding compound not only played an outstanding role in these early pioneering achievements of industrial design. It was also important in giving electric switches, lamps and lampshades new properties. An article from 1932 on "The importance of the white moulding compound 'Resopal' for the design of modern switching devices and light fittings"[42] stresses that "the new moulding compound Resopal [brings] significant progress [in that] it is now possible on the one hand to manufacture switching devices that were previously restricted to dark shades in bright colours and on the other hand to also make lampshades and bell-shaped light fittings in this material due to their transparent quality". And it went on: "Decidedly dark tones are particularly noticeable in bright rooms. The combination of several switching devices, which is necessary today, creates sizeable areas that are perceived as a disturbance as soon as they appear on a light-coloured background". "Particularly in cases where technical considerations make surface-mounted apparatus advisable, (...) greater colour choice is desirable". Today, we find this kind of interior design question completely normal, but back then it was being asked for the first time.

From 1930, Brown, Boveri & Cie. AG Mannheim had a majority and from 1934 a 100 % shareholding in H. Römmler AG, who from 1930 onwards made lamps and lampshades from translucent ivory-coloured, crème, pink or grey marbled Resopal moulding compound for BBC subsidiary Stotz-Kontakt.)[12,43] All the mouldings mentioned so far were only produced in the Römmler factories themselves for the three years from 1930

Christian Dell: Butterdosen, Senfgefäße, Zierdose und Puderdose aus dem Katalog
Resopal Kerit, 1935
Christian Dell: butter dishes, mustard pots and decorative and powder boxes from
the Resopal Kerit catalogue, 1935

Erwägungen heraus Aufputzapparate zur Ausführung empfehlen, (…) ist
eine größere Auswahl in der Farbgebung erwünscht«. Solche innenar-
chitektonischen Fragen mögen uns heute selbstverständlich sein, wurden aber damals zum ersten Male gestellt.
Ab 1930 presste die H. Römmler AG, die seit diesem Jahr mehrheitlich und
ab 1934 zu 100 Prozent zur Brown, Boveri & Cie. AG Mannheim gehörte,
Leuchten und Lampenschirme aus durchscheinender elfenbein-, creme-,
roséfarbener oder grau marmorierter Resopal-Formmasse für die BBC-
Tochter Stotz-Kontakt .[12, 43] Alle bisher genannten Formpressteile wurden
nur drei Jahre, das heißt von 1930 bis 1933 aus Resopal, das in den Römmler-
Werken selbst hergestellt wurde, produziert. Danach musste man bis zum
Ende der zivilen Produktion 1939 eingekauftes Material verwenden. Die
vollständig selbstproduzierten Artikel sind also relativ selten. Wieso?
Sofort nach der Entwicklung von Alboresin/Resopal durch die H. Römmler AG entbrannte ein heftiger Patentstreit mit der damaligen Rheinisch-
Westfälischen Sprengstoff AG in Troisdorf, die bald darauf in die Dynamit AG, vormals Alfred Nobel & Co. als Teil der IG Farbenindustrie AG,
überging. In Troisdorf hatte man gleichzeitig zu Resopal den Aminoplast
Pollopas entwickelt, einen Namen der auf den ursprünglichen Patentin-
haber der Harnstoffharz-Herstellung aus den Jahren 1921/22 – Fritz Pollack – zurückgeht. Der Prozess wurde erbittert lange geführt und ging
durch drei Instanzen. Der Streitwert betrug die damals außerordentliche
Summe von 1 Million Reichsmark.[44] Schließlich verzichtete Römmler in
einem Vertrag von 1933[45] darauf, Formpressmassen aus Aminoplasten/
Harnstoffharzen zur Verwendung in Deutschland herzustellen und gab
den Vertrieb im Deutschen Reich auf. Auslandsaktivitäten waren davon
nicht betroffen. Im Gegenzug gab die Dynamit AG die eigene Herstellung von Schichtpressstoffen und geschichtetem Plattenmaterial auf
bzw. bezahlte der H. Römmler AG Lizenzgebühren hierfür. Römmler verarbeitete ungefähr ab 1933 als Harnstoff-Formpressmasse nur noch das
Pollopas der Dynamit AG aus Troisdorf. Ansonsten machte man aus der
Not eine Tugend, konzentrierte sich mehr und mehr auf Resopal-Schicht-
pressstoffe und verteidigte bzw. festigte seine führende Stellung auf diesem Gebiet. Daher beginnt seit 1933 Resopal zur anfangs genannten all-
gemeinen Gattungsbezeichnung für Schichtpressstoffe zu werden.

to 1933, of Resopal, which was made in those same factories. After that,
during the remaining six years until the end of civilian production in 1939
they had to use bought-in material. All the articles are therefore relatively seldom. Why was that?
Immediately after Alboresin/Resopal had been developed by H. Römmler
AG, a vehement patent dispute broke out with a company that was at
the time called Rheinisch-Westfälische Sprengstoff AG in Troisdorf, and
was soon after acquired by Dynamit AG, formerly Alfred Nobel & Co. as
part of IG Farbenindustrie AG. At the same time as Römmler were developing Resopal, the company in Troisdorf had developed an aminoplast
named Pollopas, a name that stems from the original holder of the patent on urea resin manufacture from 1921/22 – Fritz Pollack. The legal dispute was long and bitter and went to appeal twice. The amount involved
was 1 million Reich marks, which was a very high sum at the time.[44] In
the end, Römmler signed a contract in 1933[45] agreeing not to manufacture moulding compounds of aminoplast/urea resins for use in Germany and ceased sales in the German Reich. Foreign operations were not
affected. In return, Dynamit AG agreed not to manufacture its own laminates and laminated panels or paid Römmler AG license fees to do so.
From 1933 onwards, Römmler processed Pollopas, a urea moulding compound made by Dynamit AG in Troisdorf. Other than that, they made
a virtue of necessity, concentrated increasingly on Resopal laminates
and defended and consolidated their leading position in this field. This
explains why Resopal began in 1933 to become the generic term for laminates, described at the beginning of this article.
Moulded articles bearing the brand name Resopal were sold until 1939,
as can be seen in adverts in *Schaulade*.[46] With the outbreak of war in
1939 civilian use of aminoplasts was generally banned.[47] Like everywhere
else, factory production switched for the most part to armaments (sea
mines, aeroplane parts, V-weapons (doodlebugs and V-2 rockets) to mention just a few).[40, 48]
In 1945 large parts of H. Römmler AG's factory in Spremberg were
destroyed.[49] The majority of the surviving machinery that was still usa-
ble was disassembled and up to 1946 transported to the Soviet Union.
Dell's phenoplast desk lamp managed to survive the destruction and dis-

Christian Dell: »Wunderkanne« aus dem Katalog
Resopal Kerit, 1935
Christian Dell: "magic coffeepot" from the Resopal
Kerit catalogue, 1935

Christian Dell: Sportservice aus Kerit
Christian Dell: stacking picnic set in Kerit

Formpressartikel der Marke Resopal wurden noch bis 1939 vertrieben, wie aus Anzeigen in der *Schaulade* hervorgeht.[46] Mit Kriegsbeginn 1939 wurde die zivile Verwendung von Aminoplasten verboten[47] und die Werksproduktion – wie überall – weitgehend auf Rüstungsproduktion (Seeminen, Flugzeugteile, V-Waffen u. v. a.) umgestellt.[40,48]

1945 wurden große Teile der Werksanlagen der H. Römmler AG in Spremberg zerstört.[49] Das Gros der noch übrig gebliebenen, verwertbaren Anlagen wurde bis 1946 demontiert und in die Sowjetunion abtransportiert.

Die Dell'sche Phenoplast-Schreibtischleuchte überlebte auf ihre Weise die Zerstörung und Demontage in Spremberg. Eine in der äußeren Form praktisch identische russische Version wurde in Orechowo-Sujewo, einer Stadt 80 km östlich von Moskau, ab etwa 1946/47 in den dortigen Karbolit-Werken auf augenscheinlich Original-Presswerkzeugen aus Spremberg gefertigt.[50] In der Sowjetunion avancierte die Dell'sche Phenoplast-Leuchte zur Büroleuchte schlechthin und ist bis heute noch zahlreich verbreitet.

Im Osten Deutschlands gelang es 1946 einigen Spremberger Werksangehörigen, die sowjetische Erlaubnis für den Aufbau eines kleinen Pressstoffwerkes aus verbliebenen Restbeständen des alten Werkes zu erhalten. Aber bereits 1948 wurde der Betrieb als VEB Plasta enteignet und verstaatlicht. Zehn Jahre später wurde er in VEB Pressstoffwerke Spremberg umbenannt, der 1970 in das Kombinat Duroplast-Halbzeuge eingegliedert wurde. Zwar nahm man in Spremberg um 1955, nach einer Pause von insgesamt 16 Jahren, die Produktion von Formpressartikeln für den Haushalt wieder auf, allerdings waren die alten Pressformen zerstört oder demontiert und abtransportiert worden. Der Formpressstoff bestand nun aus den neu entwickelten Melaminharzen, die bis heute für Haushaltsartikel verwendet werden. Die Produktion von Sprelacart-Schichtstoffen (*Sprela* = Spremberger Laminat) wurde ebenfalls 1955 aufgenommen. Ab Ende der sechziger Jahre konzentrierte man sich hauptsächlich auf diese Schichtstoffe, aber auch auf Gleitlager[51] und stellte die Produktion von Meladur-Formpressartikeln 1972 endgültig ein bzw. verlagerte sie nach Bulgarien.[52]

Im Westen gründete nach Kriegsende die Firmeneigentümerin BBC mit ehemaligen Spremberger Werksangehörigen in Groß-Umstadt eine

assembly activities in Spremberg in its own particular way. From about 1946/47, a Russian version of the lamp, externally practically identical to the original, was manufactured in the Carbolite Factory in Orekhovo-Zuevo, a city 50 miles east of Moscow, using apparently original moulding presses from Spremberg.[50] Dell's phenoplast desk lamp became the universally used office lamp in the Soviet Union and is still in widespread use today.

In East Germany a number of employees at the Spremberg works managed to get permission from the Soviets in 1946 to set up a small factory for moulded materials from remnants of old machinery that were lying around in the rubble of the old factory. In 1948 the business was disappropriated and nationalized and became known as VEB Plasta. In 1958 it was renamed VEB-Pressstoffwerke Spremberg and in 1970 was incorporated into the combine Duroplast-Halbzeuge. Around 1955 production of moulded household articles recommenced in Spremberg after a break of 16 years, but the old moulds had either been destroyed or disassembled and transported elsewhere. The moulding compound now consisted of the newly developed melamine resins, which are still in general use today for household articles. The production of Sprelacart laminates (*Sprela* = Spremberg laminate) also started up in 1955. From the late 1960s, they concentrated mainly on these laminates, but also on sliding bearings[51] and in 1972, production of Meladur moulded articles finally ceased, or rather moved to Bulgaria.[52]

After the War, the owner of the Spremberg-based company BBC joined forces with a number of its former employees in Gross-Umstadt in West Germany to found an insulation materials division, which continued to operate from 1951 as H. Römmler GmbH.[53] Here too production of moulded articles from melamine resins resumed in the 1950s and 60s, but played an ever decreasing role in the company's range of products.

In 1950 Jupp Ernst was commissioned to design the new logo for Resopal that would become so important for the company's future image. The employees in Gross-Umstadt affectionately called it "the bow tie".[41,48,54] In his designs for the advertising, trade show materials and decor development, Jupp Ernst created a particular appearance for the laminated products. It is thanks to him that the brand Resopal definitively became

Abteilung für Isolierstoffe,[1] die ab 1951 als H. Römmler GmbH fortge-führt wurde.[53] Auch hier begann man in den fünfziger Jahren die Pro-duktion von Formpressartikeln aus Melaminharz, sie spielte aber augen-scheinlich keine große Rolle mehr im Firmenprogramm.[48]

Jupp Ernst entwarf 1950 für Resopal ein für das zukünftige Auftreten der Firma sehr wichtiges neues Logo, intern liebevoll »die Fliege« genannt.[41, 48,54] Darüber hinaus bestimmte er durch die Gestaltung von Werbung, von Messeauftritten und von Produktdekor das Erscheinungsbild der Lami-nat-Erzeugnisse aus Groß-Umstadt auf Jahre. Durch ihn wurde Resopal zur prägenden Innenraumgestaltung und endgültig zum Synonym für kunststoffbeschichtete Platten.[54] 1971 wurde der so erfolgreiche Produkt-name Resopal daher als Name für die gesamte Firma eingeführt.[55]

Durch die im Jahre 1933 von der Konkurrenz erzwungene Aufgabe von Produktion und Verwendung der Arminoplast-Formpressmasse, das generelle Ende der zivilen Kunststoffproduktion in Deutschland ab Kriegsbeginn 1939, die Zerstörung der Fabrikationsanlagen bei Kriegs-ende, die sowjetische Demontage der intakt gebliebenen Anlagen sowie Maschinen und Presswerkzeuge und schließlich durch den getrennten Wiederaufbau unter jeweils völlig anderen Bedingungen im zweigeteil-ten Deutschland verlor sich die Kenntnis, dass das Harnstoffharz Reso-pal ursprünglich einmal weit mehr als Laminat war. Christian Dell muss-te als ehemaliger Bauhäusler auf Druck der Nazis seine Position an der Frankfurter Kunstschule aufgeben, was er wohl nie verwunden hat.[56] Walter Gropius bot ihm mehrmals an, wie andere ehemalige Bauhaus-leute in die USA zu gehen.[56] Dell ging ins innere Exil und blieb. Nach dem Kriege – 1945 war er 52 Jahre alt – nahm er keine feste Tätigkeit in Lehre oder Industrie mehr auf. Er kehrte zu seinen Anfängen zurück, fertigte Silberwaren und eröffnete 1948 ein Juweliergeschäft in Wiesbaden, das er bis 1955 betrieb. Er starb völlig zurückgezogen im Jahre 1974.[57] All dies zusammengenommen bewirkte, dass Christian Dells Gestaltung von Artikeln aus Formpressstoffen für die H. Römmler AG in Spremberg und ihre Bedeutung für die Designgeschichte bislang völlig unzureichend gewürdigt wurde.

So wird selbst heute auch unter Experten noch weithin der Standpunkt vertreten, dass die Geschichte des Kunststoffdesigns in Deutschland

a synonym for plastic laminates, which dominated interior design for years to come. In 1971, the name of the product that had been so suc-cessful was chosen as the name for the entire company.[55]

In being forced by competitors to abandon the production and use of Resopal as a moulding compound in 1933, the general end of civilian plastics production in Germany from the outbreak of war in 1939, the destruction of the production machinery at the end of the war along with Soviet disassembly of any plant, machinery and moulding tools that had remained intact, and finally the separate reconstruction of the com-pany, albeit under totally different conditions in the two halves of Ger-many, led to general ignorance that the urea resin the consumer knew as Resopal was originally far more than just a laminate. Christian Dell, whose name was never cited as a designer for Römmler AG in the 1930s, was, as a former Bauhaus member, forced by the Nazis to give up his post at the Frankfurt Art School, something he never got over.[56] Walter Gro-pius offered him on several occasions to go to the USA as other Bauhaus teachers had already done. Dell went into inner exile and stayed in the country. He was 52 when the war ended in 1945 and never accepted a fur-ther fixed post in academia or industry. He returned to his origins, made silverware and in 1948 opened a jewellery shop, which he ran until 1955. After that he lived a private life in Wiesbaden until his death in 1974.[57] The combination of these factors meant that Christian Dell's designs for moulded articles made by H. Römmler AG in Spremberg and their impor-tance in the history of design has never been properly recognised and is therefore not generally known by the public.

Thus, even today, experts continue to believe that the history of plas-tic design in Germany did not begin until the late 1950s to early 1960s. It was not until then, they claim, that plastics became known enough for "a great deal of valuable experience to be acquired in this field" and thus for better design possibilities and (...) more economic production. In order to "open up the market for plastic products," they maintain the idea at the time was that, " (...) design also needed to be included in the marketing process".[58] This is true of thermoplastics which were mass-pro-duced for the first time after the war, but it completely ignores the syn-thetic resin moulding compounds of the pre-war era. Wilhelm Wagen-

Nachttischleuchten aus dem Prospekt
von Stotz-Kontakt, 1931
Bedside lamps from the Stotz-Kontakt
brochure, 1931

Pendelleuchten aus dem Prospekt von
Stotz-Kontakt, 1931
Pendant lamps from the Stotz-Kontakt
brochure, 1931

Ende der fünfziger bis Anfang der sechziger Jahre beginne. Aus dieser Zeit stammt die Meinung: »Während noch vor zehn Jahren über Kunststoffe wenig bekannt war, wurden in den letzten Jahren viele wertvolle Erfahrungen auf diesem Gebiete gesammelt«, um »bessere Möglichkeiten der Formgebung und (...) eine wirtschaftlichere Produktion« zu erlauben. Zur »Erschließung des Marktes für Kunststoffprodukte« hätte es in dieser Zeit gegolten, » (...) auch das Design in diesen Vermarktungsprozess mit einzubeziehen«.[58] Dies trifft zwar auf die nach dem Krieg erstmals in großen Mengen produzierten thermoplastischen Kunststoffe zu, lässt aber die Kunstharz-Formpressstoffe der Vorkriegszeit völlig außer Acht. Wilhelm Wagenfeld wird als der »erste Industriedesigner« überhaupt genannt, der zudem in den ausgehenden fünfziger Jahren »den Weg des Kunststoffdesigns mit geebnet« habe.[59] Alle diese Aussagen negieren vollständig die Rolle derjenigen – wie Christian Dell,[60, 61] Friedrich Adler,[61, 62] Walter Maria Kersting[61, 63] und später auch Hermann Gretsch[64] –, die sich ab 1929/30 mit der Formgestaltung von Kunststoffartikeln befassten. Sie sind als frühe Industriedesigner die tatsächlichen Pioniere des Kunststoffdesigns in Deutschland, die aus ganz unterschiedlichen Gründen als solche vergessen wurden.[61]

feld is cited as being the "first industrial designer", at the close of the 1950s that "played a part in smoothing the way for plastic design".[59] All these claims completely negate the role of the people who from 1930 onwards were involved in the design of plastic articles – Christian Dell[60, 61], Friedrich Adler[61, 62], Walter Maria Kersting[61, 63] and later Hermann Gretsch[64]. They are the real pioneers of plastic design in Germany, who for very different reasons were not recognised as such.[61]

Anzeige von 1930
Advertisement from 1930

KUNSTHARZPRODUKTE AUS PHENOPLAST, DEM VORGÄNGERMATERIAL VON RESOPAL

PRODUCTS MADE OF PHENOPLAST, THE SYNTHETIC RESIN FORERUNNER OF RESOPAL

Seit 1919 hatte die Pressstoffwerke H. Römmler AG in Spremberg – Mutterfirma der heutigen Resopal GmbH – als einziges deutsches Unternehmen ein Mitnutzungsrecht des Bakelit-Patents inne. Dies bedeutete einen Vorsprung bei der Produktion von Kunstharz-Pressstoffen auf Phenolbasis, die als Hartpapierplatten, Vorläufer der Resopalplatte, und als Pressartikel auf den Markt kamen.

Die ersten vollsynthetischen Kunststoffe fanden reißenden Absatz in der aufstrebenden Elektroindustrie, die diese für Isolationsteile ihrer Maschinen oder Elektroprodukte benötigte. Der Vorteil der neuen Kunststoffe war, dass sie teure Rohstoffe wie Messing und Bronze ersetzen konnten und günstigere Eigenschaften aufwiesen als natürliche Materialien.

Den privaten Verbraucher erreichten die Produkte der H. Römmler AG als moderne Gebrauchsartikel für den Haushalt und mit den neuen technischen Errungenschaften wie Telefon, Rundfunk und Auto, deren Gehäuse bzw. Armaturentafeln aus dem Römmler-Bakelit Hares bestanden, das in veränderter Form bis in die siebziger Jahre produziert wurde. Auch Bügeleisengriffe, Föhngehäuse, Kippschalter, aber vor allem die Gehäuse für Millionen Volksempfänger kamen aus Spremberg.

Ab 1933 – zu dieser Zeit hatte das Unternehmen etwa 850 Mitarbeiter – mussten aufgrund der verordneten, immensen Nachfrage von Volksempfängern permanent Anträge auf Sonntagsarbeit und Überstunden beim Gewerbeaufsichtsamt Cottbus gestellt werden. Außerdem machten ab Mitte der dreißiger Jahre die »Fertigstellung wehrwichtiger Artikel« und »größere kurzfristige Aufträge auch seitens der Heeresverwaltung die Weiterbewilligung von Sonntagsschichten erforderlich.«

Ein Gutachten, das 1934 von den Aktionären der H. Römmler AG zur Wirtschaftlichkeit des Betriebes in Auftrag gegeben wurde, kam zu dem Schluss: »Die Entwicklung der Umsätze in den verschiedenen Artikeln und in den verschiedenen Branchen beurteilen wir als außerordentlich günstig. Für die Zukunft ist bei diesem Unternehmen der seltene Fall gegeben, daß die Umsatzmöglichkeiten unübersehbar gut sind und daß es nur darauf ankommt, rechtzeitig den Bedürfnissen des Marktes mit einer entsprechend sparsamen Beeinflussung der Fabrikation nachzukommen.«

Since 1919, Pressstoffwerke H. Römmler AG in Spremberg – the parent company of today's Resopal GmbH – has been the only German company to have the rights to use the Bakelite patent. This put them one step ahead of the competition in manufacturing materials based on phenolic resin marketed as laminated papers, the forerunner of the Resopal laminate, and as moulded articles.

The first fully synthetic plastic recorded phenomenal sales in the burgeoning electrical industry, which needed insulation material for its machines and electrical appliances. The advantage of the new plastics was that they could be used instead of expensive raw materials such as brass or bronze, and had better properties than natural materials.

The products made by H. Römmler AG reached the consumer as modern everyday articles and in new technological achievements such as telephones, wireless sets and cars, whose casings and instrument panels were made of their version of Bakelite known as Hares. It continued to be produced, albeit in a modified form, until the seventies. Handles for irons, hairdryer casings, electrical switches and above all the housing for the *Volksempfänger* wireless set came from Spremberg.

From 1933 onwards – at the time the company had a workforce of about 850 – the immense demand for the *Volksempfänger*, which every German family was obliged by the government to own, meant that the company constantly had to apply to the factories inspectorate in Cottbus for permission to work overtime or on Sundays. Furthermore, from the mid-1930s, the "manufacture of articles of military importance" and "large short-notice orders from the military administration made it necessary to continue issuing permits for Sunday shifts."

A report commissioned in 1934 by the shareholders of H. Römmler AG on the company's economic feasibility reached the following conclusion: "We consider the growth in sales for the different articles and in the different sectors extremely positive. For the future of the company, the rare scenario is true that sales possibilities are unmistakably good and it is merely a matter of supplying the needs of the market promptly and ensuring that production remains economical."

hRs
Schnallen-Isolatoren Din 3170
sind Abspann-Isolatoren besonders hoher Zugfestigkeit.
Sie bestehen aus einem Stahlrahmen, der mit hochwer-
tigem Isoliermaterial umpreßt ist. Durch besondere Anordnung lassen sich Verbindungen herstellen, die für höhere
elektrische und mechanische Belastungen geeignet sind.
Anwendungsgebiete: Straßenbahnen, Grubenbahnen,
Verladebetriebe, Transportanlagen, Hebezeuge usw.

■ **JUPP ERNST: PROSPEKTE FÜR DIE H. RÖMMLER AG, UM 1938**

In einem räumlichen Bildaufbau werden frei gestellte technische Produkte zu zeichenhaften Gebilden und damit zur Attraktion. Die Werbegrafik entspricht ganz der Forderung des Werkbunds und des Bauhauses, die Information auf »die Eindeutigkeit des Wirklichen, Wahren in der Alltagssituation« zu reduzieren und auf den »Ballast falscher Pracht« zu verzichten. Jupp Ernst wird nach dem Krieg die gesamte Werbung für Resopal übernehmen.

■ **JUPP ERNST: BROCHURES FOR H. RÖMMLER AG, AROUND 1938**

In a three-dimensional photo composition, freely arranged technical products have become symbolic structures and thus an eyecatcher. The advertising graphics fully meet the requirement of the Werkbund and Bauhaus that information be reduced to the "unequivocal quality of what is true and authentic in the everyday situation" and should dispense with the "ballast of false grandeur." After the War, Jupp Ernst would take over all Resopal's advertising.

1

2

■ KUNSTHARZPRODUKTE DER H. RÖMMLER AG, NACH 1933
Kunststoffe erfuhren als Werkstoffe des Vierjahresplans besondere
Förderung durch die Nationalsozialisten für deren Kriegswirtschaft.
1 Propagandaplakat 2 Warnschild aus Resopal 3 Prospekt von Jupp
Ernst 4 Handräder in einem deutschen U-Boot

■ SYNTHETIC RESIN PRODUCTS MADE BY H. RÖMMLER AG, POST-1933
Plastics were amongst the materials particularly promoted in the four-
year plan the Nazis drew up for their war economy.
1 Propaganda poster 2 Warning sign in Resopal 3 Brochure by Jupp
Ernst 4 Handwheels in a German U-boat

3

4

BEHÄLTER FÜR DOPPELFERNROHR 6 × 30, 1942

Welche deutschen Unternehmen an der Herstellung von Rüstungsge-
genständen beteiligt waren, lässt sich aufgrund der Kennzeichnungspflicht der nationalsozialistischen Behörden leicht nachweisen, so auch
bei diesem Objekt. Der eingeprägte Code »ehe« auf dem Behälter des
Fernglases besagt, dass dieser von der H. Römmler AG hergestellt wurde.
Der Verschluss des Behälters, der mit »fm 42« markiert ist, verweist darauf, dass die Firma Rudolf Lang in Brandenburg/Havel dieses Produkt 1942
hergestellt hat. Der Firmencode »blc« auf der Vorderseite des Feldste-
chers gibt an, dass die optischen Gläser von Carl Zeiss Jena stammen.
Noch vorhandene Unterlagen beweisen direkte Kontakte der H. Römmler AG zum Oberkommando des Heeres, in dessen Auftrag Behälter für
Rüstungsgegenstände hergestellt wurden. Von diesem Amt wurde auch
verfügt, welche Pressmassen verwendet werden durften. Ein anderes
Dokument besagt, dass die H. Römmler AG Hartgewebebuchsen für Pan-
zerkraftwagen fertigen sollte. Von 1940 bis 1944 liegen Umsatzzahlen für
Lieferungen von Pressmassen an eine Vielzahl von Unternehmen vor, so
auch für die Firma »Sils & Co, Fröndenberg, Werk Thorn, Auschwitz«.

CASE FOR 6 × 30 BINOCULARS, 1942

The mandatory labelling imposed by the Nazi authorities makes it easy
to identify which German companies were involved in the production of
armaments. Thus, in the case of this article, the code "ehe" stamped on
the binoculars case indicates that it was made by H. Römmler AG. The
cap, which bears the mark "fm 42," shows that the company that manufactured it in 1942 was Rudolf Lang in Brandenburg/Havel. The company code "blc" on the front of the field glasses shows that the lenses were
made by Carl Zeiss Jena.
Extant records prove that there was direct contact between H. Römmler
AG and the military high command, which commissioned cases and
containers for armaments. The office of the high command also dictat-
ed which moulding compounds could be used. Another document also
stated that H. Römmler AG should manufacture laminated-fabric bearings for armoured vehicles. Sales figures for the years 1940 to 1944 show
deliveries of moulding compound to many different companies, including "Sils & Co, Fröndenberg, Werk Thorn, Auschwitz."

Stellvertretend für eine breite Produktpalette stehen hier ein elektrischer Haartrockner, eine Tischleuchte und ein Rundfunkgerät. Beim lichtdurchlässigen Schirm der als »Bauhaus-Entwurf« titulierten Lampe ist das als Marmorierung erscheinende Pressstoffgefüge gut zu erkennen. Das Radiogehäuse für den Deutschen Kleinempfänger, die sogenannte »Goebbels' Schnauze«, wurde ab 1938 produziert.

The electrical hairdryer, table lamp and wireless set depicted here represent the broad range of electrical articles in the company's range. In the translucent shade of the lamp described as a "Bauhaus design" the structure of the moulding compound, which looks like marbling, is easy to identify. The housing for the small wireless set, the *Deutscher Kleinempfänger*, known in popular parlance as "Goebbels' gob," was produced from 1938 onwards.

■ **GEBRAUCHSARTIKEL DER H. RÖMMLER AG, DREISSIGER JAHRE**

Außer Dosen jeglicher Art waren zunächst Tabletts, Aschenbecher und Untersätze aus Kunstharz-Pressstoff weit verbreitet. Sie zeichneten sich durch eine moderne Zweckform und ihren natürlichen Glanz aus. Als Vorteil wurde insbesondere die Wirtschaftlichkeit des neuen Materials betont, dessen Oberfläche nicht wie bei Holz und Metall extra behandelt werden musste. Praktischer Vorzug im täglichen Gebrauch war die leichte Reinigung.

■ **EVERYDAY ARTICLES MADE BY H. RÖMMLER AG, 1930s**

Apart from containers of all descriptions, trays, ash trays and coasters of moulded synthetic resin enjoyed widespread popularity. They were characterised by their modern functional form and natural lustre. The economic nature of the new material was emphasised as a particular benefit: unlike wood or metal it did not need any special surface finishing. Its practical advantage in everyday use was that it was easy to clean.

DOSE DER FIRMA KAFFEE HAG, 1937/1939

Die Phenoplast-Pressartikel der H. Römmler AG haben auf den Unterseiten der einzelnen Teile die Einprägung »HRS«. Außerdem finden sich hier die Nummer der verwendeten Pressform, in diesem Fall über dem Schriftzug »Kaffee HAG«, und neben dem Firmenkürzel noch ein Prüfzeichen, das den Hersteller und die genutzte Pressmasse kenntlich macht. Aufgrund der sehr großen, steigenden Anzahl von Presswerken und einer dementsprechend unübersichtlichen Marktlage, stellte sich bereits in den zwanziger Jahren dringend die Frage nach einer Qualitätskontrolle und Kennzeichnungspflicht. Der Verband Deutscher Elektrotechniker entwickelte die ersten Richtlinien zur Typisierung von Pressmassen, denn er vertrat mit der Elektroindustrie einen der Hauptabnehmer dieses Materials. 1924 wurde die Technische Vereinigung der Hersteller typisierter Pressmassen und Pressstoffe gegründet, deren Mitglieder, zu denen auch die H. Römmler AG zählte, sich einer freiwilligen Prüfung durch das Staatliche Materialprüfungsamt in Berlin-Dahlem unterzogen. In dem Prüfzeichen, das ab 1928 vergeben wurde, identifiziert ein zweistelliger Code über dem stilisierten »M« – hier die Zahl 32 – das Presswerk, die H. Römmler AG. Der Code unter dem »M« bezeichnet die Art der Pressmasse.

KAFFEE HAG JAR, 1937/1939

All the components of the phenoplast moulded articles that were made by H. Römmler AG bear the mark "HRS" on the underside. The number of the mould used is also marked in the same place, in this case above the lettering "Kaffee HAG." Next to the company's abbreviation HRS there is also a certification mark, indicating the manufacturer and the moulding compound used.

Due to the vast and ever-increasing number of moulding factories and a correspondingly confused market situation, the question of quality control and mandatory labelling arose as a matter of urgency as early as the 1920s. The German professional association of electrical engineers (VDE) developed the first guidelines on standardisation of moulding compounds, since it represented one of the main purchasers of this material, the electrical industry. In 1924, the "Technische Vereinigung der Hersteller typisierter Pressmassen und Pressstoffe", an association of manufacturers of standardised moulding materials and compounds, was founded. Its members, who included H. Römmler AG, entered a voluntary agreement to undergo review by the state materials testing agency in Dahlem, Berlin. The certification mark, which was awarded from 1928 onwards, consists of a two-digit code above a stylised "M" – in this case 32 – identifying the moulding factory, H. Römmler AG. The code beneath the "M" indicates the type of moulding compound.

■ **PROSPEKT DER H. RÖMMLER AG, UM 1930**

Die Kunststoffindustrie in Deutschland war seit Mitte der zwanziger
Jahre mit der Entwicklung neuer Harze beschäftigt, die im Gegensatz
zu den dunklen Phenoplasten auch in hellen Farben und sogar reinweiß hergestellt werden konnten. Die H. Römmler AG stellte erstmals
Haushaltsartikel aus hellfarbigem Aminoplast auf der Leipziger Frühjahrsmesse 1930 aus. Allerdings hieß der neue Pressstoff noch nicht
Resopal, sondern Alboresin.

■ **BROCHURE BY H. RÖMMLER AG, AROUND 1930**

Since the mid-1920s, the plastics industry in Germany had been working on the development of new resins, which, by contrast with the
dark phenoplasts, could also be produced in light colours or even pure
white. H. Römmler AG exhibited household articles in light-coloured
aminoplast for the first time at the Leipzig trade show in spring 1930.
However, the new moulding compound was not yet called Resopal, but
Alboresin.

Die Unangreifbarkeit der Oberfläche des Materials gegen Säfte und Säuren führte zur Herstellung formenschöner Obstschalen, Fingerschalen etc.

Als letzten Erfolg unseres Schaffens bringen wir auch Gebrauchs-Gegenstände aus der lichtbeständigen, rein weiß leuchtenden

Kunstharz-Preßmasse „Alboresin"

wofür wir Patente in allen Kultur-Staaten angemeldet haben, die vielfach bereits erteilt sind. Die isolierende Eigenschaft des Kunst-harz-Preßstoffes ließ auch die Herstellung von Zigarettenkästen ratsam erscheinen. Die in diesen Kästen aufbewahrten Rauchmittel behalten lange Zeit ihren frischen Wohlgeschmack. Praktisch und nach neuzeitlichem Geschmack geformte Aschenbecher sind wiederum leicht zu reinigen, da auch glimmendes Feuer deren mattglänzende

Oberfläche nicht zerstört. Das Herstellungsprogramm der Preßstoff-Erzeugnisse, zu dem heute auch Bieruntersätze und Teeschalen ge-hören, erfährt durch Hinzunahme weiterer Artikel eine ständige Vergrößerung. Es bedarf kaum der Erwähnung, daß alle diese Gebrauchsgegenstände sich nicht nur für den Privathaushalt, sondern auch aus wirtschaftlichen Gründen für Hotels, Gaststätten besonders eignen.

H. RÖMMLER

AKTIENGESELLSCHAFT
PRESSTOFF-WERKE
SPREMBERG N. L.

Jupp Ernst: Erster Prospekt für die Anwendung von Resopalplatten im Innenausbau, um 1938
Jupp Ernst: First brochure on the use of Resopal laminates in interior finishing, around 1938

RESOPAL ALS PRESSMASSE UND SCHICHTSTOFFPLATTE
RESOPAL AS A MOULDING COMPOUND AND LAMINATE

Der Weg bis zur heutigen Resopalplatte setzte eine lange technische Entwicklung voraus.

Am 19. Dezember 1930 wurde das sogenannte Aminoplast-Patent von der H. Römmler AG eingereicht, die auf Grund dieser Erfindung als erstes Unternehmen in der Welt hellfarbige Schichtstoffplatten produzieren und unter dem Markenzeichen Resopal in den Handel bringen konnte. Die Nutzung des neuen Verfahrens begann allerdings wie im Fall der dunklen Phenoplastprodukte mit der technischen Entwicklung und der industriellen Herstellung plastischer Massen und Pressartikel. Die Schichtstoffplatte war ein Spezialfall dieser Technik.

Die ersten Platten aus Aminoplast waren transparent. Um farbige Platten herstellen zu können, war die Entwicklung spezieller Zellulose notwendig. Als die entsprechenden Spezialpapiere zur Verfügung standen, erschlossen sich für die Verwendung der Schichtstoffplatte neue Bereiche. Bei den Mehrschichtplatten, die aus farbigen Papierlagen bestanden, konnte durch Gravierung der obersten Schicht die Farbe der zweiten Schicht sichtbar gemacht werden, eine Technik, die sich vor allem für Schilder mit Schrift eignete. Über eine andere Methode, das Unterdruckverfahren, konnten Werbetafeln produziert werden, indem zum Beispiel farbige Drucke eingepresst wurden.

Ab 1938 wurden auch Platten größeren Formats und größerer Dicke hergestellt. Da dieses Verfahren aber zwei Pressvorgänge erforderte, suchte man nach einer Methode, um in nur einem Arbeitsgang das Material mit einer Aminoplast-Oberfläche und einem Kern aus Phenol-Kresol-Harz zu fertigen. Erst nachdem eine dafür geeignete Harzverbindung entwickelt worden war, stand der weiten Verbreitung der hellen Resopalplatte nichts mehr im Wege.

Noch nicht vertraut mit den neuen farbigen Möglichkeiten, wünschten viele Kunden holzähnliche Oberflächen. Nach einigen Experimenten mit Holzimitaten verwendete die H. Römmler AG schließlich Echtholztapeten einer Berliner Firma und entwickelte Oberflächen in vier verschiedenen Holzdekoren, die auch optisch den Ansprüchen der Kunden gerecht wurden. Platten dieser Art waren sogar für die Innenverkleidung der Waggons der Deutschen Reichsbahn vorgesehen. Allerdings kam es – vermutlich wegen des Kriegsausbruchs – nicht zu einer praktischen Ausführung.

What we know today as the Resopal laminate is the culmination of a long process of technical development.

On 19 December 1930, the so-called aminoplast-patent was filed by H. Römmler AG, which as a result of this invention became the first company in the world to be able to produce and market light-coloured laminates under the Resopal brand name. However, use of the new process began, as had been the case with the dark-coloured phenoplast products, with the technical development and industrial manufacture of plastic compounds and moulded articles. The laminated board represented a special application of this technology.

The first aminoplast laminates were transparent. To be able to make coloured laminates it was necessary to develop a particular kind of cellulose. When the specialty papers became available, new possibilities for using laminates opened up. In the case of multi-layered laminates, which consisted of layers of coloured paper, the top layer could be engraved to reveal the colour of the layer beneath, a technique that was particularly suitable for signs containing text. A different method could be used to produce advertising placards, for example, whereby coloured prints were pressed between the core sheets and the upper layer of protective resin.

From 1938, the laminates also began to be made in larger formats and greater thicknesses. However, since this process required two pressing procedures, attempts were made to find a method by which an aminoplast surface finish and a core of paper impregnated with phenolic and cresylic resins could be fused together in a single pass. It was not until a resin compound that was suitable for this had been cultivated that the development of the light-coloured Resopal laminate could advance unhindered.

Not yet familiar with the new coloured options, many customers still wanted wood-effect surfaces. After a number of experiments with wood imitations, H. Römmler AG finally used only wood-veneer wallpaper made by a Berlin-based company and then developed surface finishes in four different wood designs, which also met their customers' visual demands. Laminates of this type were even due to be used for the interior panelling of Deutsche Reichsbahn carriages, but they were never actually made – presumably due to the outbreak of war.

■ CHRISTIAN DELL: RESOPALGESCHIRR, VOR 1935

Resopal-Pressstoffartikel der H. Römmler AG wurden von der Plastica GmbH, einer Berliner Verkaufsgemeinschaft für Pressartikel vertrieben. Im Katalog für Wiederverkäufer wurden auch die hier abgebildeten Geschirrteile – mit Ausnahme der Tasse mit Silberhalterung und Ebenholzgriff – veröffentlicht. Sensationell war die »Sport-« oder »Wunderkanne«, weil sie sechs Tassen, einschließlich Untertassen sowie Sahnegießer und Zuckerdose aufnehmen konnte.

■ CHRISTIAN DELL: RESOPAL TABLEWARE, PRE-1935

Moulded articles in Resopal produced by H. Römmler AG were marketed by Plastica GmbH, a Berlin-based sales association specialising in moulded articles. With the exception of the cup with a silver holder and ebony handle, the pieces depicted here were also published in the resellers' catalogue. The "sports" or "magic coffee pot" was sensational in that it could hold six cups and saucers, a cream jug and sugar bowl.

Die Firma Venditor in Troisdorf (bei Köln) war die Verkaufsgesellschaft für Kunststoffe innerhalb der Dynamit Aktiengesellschaft (D.A.G.), vormals Alfred Nobel & Co. Die D.A.G. war eine der größten Konkurrenten der H. Römmler AG. Die im Werbeprospekt abgebildeten Haushaltsgegenstände bestehen aus dem resopalähnlichen Material Ultrapas, die beiden Tassen aus Pollopas.

Venditor, a company based in Troisdorf (near Cologne), was the sales organisation for plastics within Dynamit Aktiengesellschaft (D.A.G.), formerly Alfred Nobel & Co. D.A.G. was one of H. Römmler AG's major competitors. The household articles depicted in the advertising brochure are made of Ultrapas, a material similar to Resopal, the two cups are made of Pollopas.

Die Schwarzweiß-Abbildungen aus einem zeitgenössischen Prospekt zeigen wichtige frühe Anwendungsbereiche der Kunststoffplatte: Deckenverkleidung einer Straßenbahn, Duschwände für eine öffentliche Badeanstalt und Belag eines Schreibtisches, auf dem auch ein Telefon steht, dessen Gehäuse ebenfalls von der H. Römmler AG stammte. Weiter verbreitet waren vor dem Zweiten Weltkrieg allerdings Warn-, Hinweis- und Werbeschilder aus Resopal.

The black-and-white illustrations from a contemporary brochure show important early applications of the laminated panel: the ceiling of a tram, the walls of showers in public baths, and the surface finish of a desk with a telephone, the housing for which was also manufactured by H. Römmler AG. However, the products in Resopal that were more widely used before the Second World War were warning, information and advertising signs.

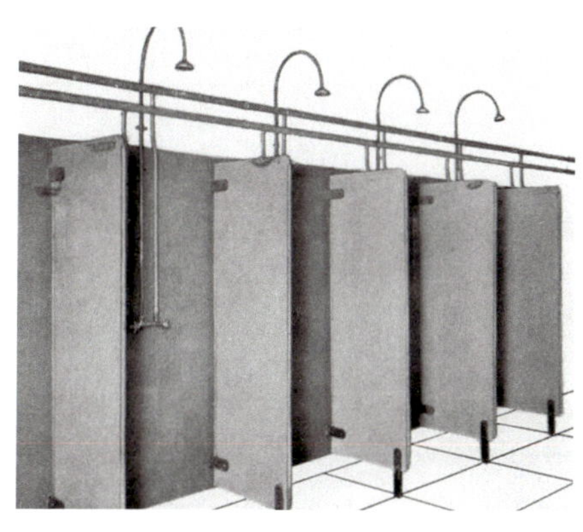

■ SCHILDER AUS RESOPAL, DREISSIGER JAHRE

Das obige Werbeplakat wurde im Unterdruckverfahren hergestellt. Vielleicht konnte es sogar hinterleuchtet werden, ein besonders gepriesener Vorzug des Materials. Der Prospekt, der von Jupp Ernst gestaltet wurde, wirbt für eine zweite Technik, bei der die in die Platte eingravierte Schrift in der Farbe der unteren Schicht erscheint. Ein Beispiel dafür ist das Schild der Peenemünder Bahn von 1942.

■ SIGNS IN RESOPAL, 1930s

The advertising poster above was produced using a technique whereby the design is pressed between the core sheets and the upper layer of protective resin. It may have also been backlit, which was one of the material´s most highly valued qualities. The brochure, which was designed by Jupp Ernst, is advertising a second technique, in which the lettering that has been engraved into the board appears in the colour of the lower layer of paper. An example of this is the sign for the Peenemünde railway of 1942.

SCHILD AUS DER S-BAHN BERLIN, 1938

Mit der Bauart 1937 begann der Einsatz von Resopal bei der Berliner S-Bahn. Die aus dem neuen Werkstoff produzierten Innenschilder für den Fahrgastraum ersetzten die bisher verwendeten Fensterschilder aus Blech, die emaillierten Raucher- und Nichtraucherschilder sowie die in Glas gerahmten Papptafeln mit Verkehrsregeln. Da man zunächst noch keine Erfahrungen mit Resopal hatte, beobachtete die Reichsbahn die Beständigkeit der Schilder genau. Nach fünfjähriger Einsatzzeit konnte festgestellt werden, dass der weiße Farbton lediglich leicht vergilbt war. Die bisher verwendeten Papptafeln hingegen mussten spätestens nach drei Jahren ersetzt werden. Die Herstellung einer Tafel kostete zwar nur fünf Pfennig und war damit zunächst deutlich billiger als ein Resopalschild für 2,75 Mark, aber unter Berücksichtigung, dass Papptafeln noch auf Glasscheiben aufgezogen, im Rahmen verklebt werden mussten und zudem noch eine wesentlich kürzere Haltbarkeitszeit hatten, war der Einsatz von Resopalschildern deutlich ökonomischer. Unter Berücksichtigung aller Kosten waren pro Waggon für Schilder aus Resopal nur 10,31 Mark, für Schilder aus Pappe im gleichen Zeitraum 12,50 Mark aufzuwenden.

A SIGN FROM THE BERLIN S-BAHN, 1938

The Reichsbahn, who were responsible for running Berlin's public transport, began to use Resopal for the rolling stock on its suburban railway network, or S-Bahn, in 1937. The signage produced from the new material for use inside the carriages replaced the sheet metal window signs, the enamel smoking and no-smoking signs and the railway byelaws that were printed on cardboard and framed in glass. Since they had no experience of using Resopal, the Reichsbahn kept a close eye on how durable the signs were. After five years of use, the only sign of wear was that the white had yellowed slightly. By contrast, the previously used cardboard signs had to be replaced after three years at the latest. Although it cost only a few pennies to make a cardboard sign, which was significantly cheaper than a Resopal sign at 2.75 Reichsmarks, it had to be taken into account that the cardboard signs had to be mounted and framed behind glass and also lasted for a much shorter time, so that using Resopal was clearly more economical. Taking all the costs into account, Resopal signs cost only 10.31 Reichsmarks per carriage, whereas cardboard ones for the same period of time would have cost 12.50 Reichsmarks.

Gerda Breuer

»VERTREIBUNG AUS DEM ERSTEN PARADIES« – KUNSTSTOFF IN DER BUNDESREPUBLIK DER FÜNFZIGER JAHRE IM KONTEXT VON DESIGNDISKURSEN

"EXPULSION FROM THE PRIMAL PARADISE" – PLASTIC IN WEST GERMAN DESIGN DISCOURSES IN THE 1950S

Anfang der fünfziger Jahre flammte in den Designdiskursen der BRD die nahezu ein Jahrhundert während Diskussion über die Verwendung von Kunststoffen erneut auf. Nachdem es bereits in den zwanziger und dreißiger Jahren in Ländern wie den USA und auch in Deutschland zum Durchbruch in der Gestaltung des vergleichsweise jungen Materials gekommen war, wurde vor allem von Vertretern der »Guten Form« in der Nachkriegsdekade das Für und Wider der Verwendung noch einmal grundsätzlich debattiert.

War unmittelbar nach dem Krieg die Linderung der Not von einem Pragmatismus bestimmt, der die in den Diskursen der Reformbewegungen des Kunstgewerbes ausgeprägte ethische Grundhaltung der Realität der Zeit anpasste, begannen schon bald die Auseinandersetzungen über eine Neufassung des »Zeitgemäßen« in der Formgebung, insbesondere in der industriellen Formgebung. »Jede neue Zeit fordert einen neuen Menschen« schrieb der Architekt und Werkbündler Otto Bartning 1950. »Und wenn wir nicht im tiefsten Grunde spürten, dass wir vor einer neuen Zeit stehn, brächten wir nicht den Mut auf, jeden Morgen aus unsern Trümmern und Sorgen aufzustehn. Der nur rückwärts gerichtete Wunsch, das Verlorene und Vergangene (mehr oder weniger kümmerlich) wiederherzustellen, gibt zwar allenfalls den traurigen, mit Ressentiments beladenen Mut der Verzweiflung. Wir aber brauchen den frohen Mut zum Neuen, zum Einfachen, Aufrichtigen und Gesunden, kurz den Mut zum Glück und zur Freiheit.«[1]

Vor allem in den Kreisen des Deutschen Werkbundes, der sich 1953 offiziell als Verein wiedergründete, nachdem er im »Dritten Reich« von den Nationalsozialisten aufgelöst worden war, spielten Kunststoffe als neues, zukunftsträchtiges Material eine zentrale Rolle. Zwar waren sie durch die 1907 gegründete Vereinigung nicht pauschal abgelehnt worden, aber doch in die Nähe der Scheinhaftigkeit einer bürgerlichen Kultur und seiner Kunstgewerbeindustrie eingeordnet worden: »Solche Produkte sind gegen die Wahrhaftigkeit und Ehrlichkeit, wirken in gleicher Weise unsittlich und unästhetisch«, hieß es im *Deutschen Warenbuch*, das der Werkbund 1915 als Orientierung für Entwerfer, Produzenten und Konsumenten herausgab. Aber er benannte auch Ausnahmen, die unter bestimmten Bedingungen akzeptiert werden konnten: »Am meisten befriedigen Neubildungen, die wir nicht direkt aus der Natur gewinnen, wie z. B. Celluloid. Nur dürfen sie in der künstlerischen Behandlung nicht verwandte Naturstoffe, z. B. Elfenbein oder Horn, vortäuschen wollen.«[2] In der Tat wandten sich Werkbündler und später auch Bauhäusler einigen Kunststoffen zu, was im Besonderen für Linoleum galt, doch dürfen diese Ausnahmen nicht darüber hinwegtäuschen, dass der synthetische Stoff mehr oder weniger ignoriert wurde.[3]

Nach Kriegsende stellte sich erneut die Frage, welche Haltung man der relativ jungen kunststoffverarbeitenden Industrie gegenüber einnehmen sollte, zumal Kunststoff sich in einer der Siegernationen, den USA, seit den dreißiger Jahren großer Beliebtheit erfreute. In Designdiskursen, das heißt in den einschlägigen Gazetten, wurde die Verwendung von Plastik zum Gegenstand einer lebhaften Auseinandersetzung. Flankiert wurde dieses neue Interesse für das äußerst formbare Material von der kunststoffverarbeitenden Industrie selbst, die Anschluss an internationale Entwicklungen suchte. In der deutschen Designgeschichte gab es in Gestaltungsfragen fast keine Vorbilder, die für die Unternehmen Orientierung hätten bieten können. Die breiten Einsatzmöglichkeiten des Kunststoffs, besonders für Böden, Möbel und Geräte, die bis dato aus traditionellen Materialien bestanden, stellten Herausforderungen für die Gestaltung der Form dar, wollte man das Material nicht nur auf die Rolle eines Ersatzstoffes reduzieren.

At the beginning of the 1950s the debate on the use of synthetic materials that has now lasted for almost a century was reignited in design discourses in West Germany. After a design breakthrough had been achieved with new materials back in the twenties and thirties in countries like the USA, and also in Germany, the decade after the Second World War saw the advocates of "good design" debating the fundamental pros and cons of their use. Immediately after the War, the alleviation of dire need was coloured by a certain pragmatism, which caused the decidedly ethical fundamental attitude prevalent in the discussions of the reform movements in applied arts to adapt somewhat to the reality of the time. However, disputes over a new interpretation of "contemporary design" and, in particular, "industrial design" soon began. "Each new era requires a new human being," wrote architect and Werkbund member Otto Bartning in 1950. "And if we did not realise in the depths of our being that we are facing a new era we would not be able to summon the courage to rise every morning from the ruins and worries surrounding us. The backward-looking desire to restore (more or less sparsely) what has been lost, what is in the past, gives us at best the sad courage of despair, charged with resentment. What we need is the joyful courage to embrace the new, the simple, the upright and healthy: in short, the courage to embrace happiness and freedom."[1]

Above all in Deutscher Werkbund circles, an organisation which, having been forced by the Nazis to disband during the Third Reich, had been refounded in the early post-War years in the form of individual groups and regional organisations and then officially as an association in 1953, plastics played a central role as "new" materials of the future. Although they had not been rejected out of hand by the association that was founded in 1907, they had nevertheless been classified as approaching the fake nature of the bourgeois culture and its arts and crafts industry: "Products of this kind are contrary to authenticity and truthfulness, and as such they are unseemly and unaesthetic in appearance," declared the *Deutsches Warenbuch*, a handbook published by the Werkbund in 1915 as a guidance document for designers, manufacturers and consumers alike. But it did also name exceptions that might be acceptable under certain conditions: "Most satisfactory are new inventions that we do not acquire directly from nature, such as celluloid, for example. But, in the way they are artistically treated, they must not simulate natural materials, such as ivory or horn."[2] In fact, Werkbund, and later Bauhaus, members had used some synthetic materials in the heroic years of Modernism. This was particularly true of linoleum, but the exceptions should not hide the fact that synthetic materials were more or less ignored.[3]

After the War, the question of what attitude should be adopted towards the relatively new plastics processing industry resurfaced, particularly as plastic had been extremely popular since the thirties in one of the countries that had won the War – the USA. In design discussions, i.e. in the relevant journals, the use of plastic became the subject of lively debate. This new interest in the highly malleable material was also espoused by the plastics processing industry itself, which was keen on connecting with international trends. There were virtually no design exemplars in the history of German design that could have given companies any kind of guidance. The broad range of uses for plastics, particularly for flooring, furniture and appliances that had until then been the sole domain of traditional materials, presented challenging design questions that had to be addressed, if it was not simply to be confined to the role of a substitute material.

Immediately after the War, a number of designers in the USA had helped this synthetic material achieve a breakthrough. The most famous exam-

Ausstellung »Organic Design in Home Furnishings«, Titelseite des Katalogs, 1941
Exhibition "Organic Design in Home Furnishings," catalogue cover, 1941

Möbel von Charles Eames und Eero Saarinen in der Ausstellung »Organic Design
in Home Furnishings«, The Museum of Modern Art, New York, 1941
Furniture by Charles Eames and Eero Saarinen in the exhibition "Organic Design
in Home Furnishings," The Museum of Modern Art, New York, 1941

Nun hatten in den USA einige Designer dem synthetischen Material schon unmittelbar nach Kriegsende zum Durchbruch verholfen. Zu den bekanntesten Beispielen gehörte der 1946/47 von Eero Saarinen entworfene »Womb Chair« mit einer Sitzschale aus verformtem Kunststoff, über die er allerdings noch eine mit einem Stoff bezogene Schaumgummipolsterung zog. Dann aber spielten Ray und Charles Eames bei einer Gruppe von Sesseln bzw. Stühlen mit Sitzschalen aus fiberglasverstärktem Polyester, den sogenannten »Plastic-Armchairs«, mit dem ästhetischen Reiz des Materials. Beide Möbel kamen dem formalen Anspruch des organischen Designs entgegen, das sich in den vierziger Jahren dort wie in England großer Akzeptanz erfreute und 1940 im Wettbewerb »Organic Design in Home Furnishings« des Museums of Modern Art in New York breite öffentliche Beachtung fand. Auf dieser Ausstellung wurde auch der unter dem Namen »La Chaise« bekannte, nicht prämierte edle Sessel aus Hartgummi und fiberglasverstärktem Kunststoff des Ehepaars Eames präsentiert, der zum Inbegriff des organischen Designs dieser Zeit wurde.[4] Hier wurde, wie auch bei vielen anderen Beispielen, Kunststoff nicht nur von anerkannten Designern in einem der renommiertesten Kunstmuseen der Welt präsentiert, es wurde auch in die Nähe der zeitgenössischen Kunst eines Henry Moore, Jean Arp und Gaston Lachaise gerückt, nach dem das Möbel von den Eames benannt war.

Auch im Osten Deutschlands wurde bald nach der Gründung der DDR das »neue« Material als Herausforderung für Massenproduktion und Design entdeckt. Die Verwendung und Gestaltung von Kunststoff wurden regelrecht staatlich verordnet und von den renommierten Designhochschulen in Halle, Burg Giebichenstein sowie Weimar und Berlin mitgetragen. Die »Plaste- und Elasteartikel«, häufig in selbstbewusster Schlichtheit gestaltet, überrollten Ende der fünfziger, Anfang der sechziger Jahre regelrecht das Land.

In der BRD wurde die Gestaltung von Kunststoff zunächst von vielen geradezu als unangenehm empfunden. Hans Schwippert, ab 1953 erster Vorsitzender des Deutschen Werkbundes, brachte den Gewissenskonflikt auf den Punkt, als er von der »Vertreibung aus dem ersten Paradies« sprach und vom »ersten großen Kampfabschnitt der Gestalter um das wohlgeformte und wohlanständige Ding«[5], der nun beendet sei, um in eine

ples included the "Womb Chair" designed in 1946/47 by Eero Saarinen, which had a seat shell of moulded plastic that he then upholstered with fabric-covered latex foam. Ray and Charles Eames then designed a group of chairs and armchairs with seats made of fibreglass reinforced polyester, the so-called "plastic armchairs," where they played with the aesthetic appeal of the material. Both pieces of furniture met the formal standards of organic design, which enjoyed great popularity in the forties both in Germany and Britain and met with great public acclaim in 1940 in the competition "Organic Design in Home Furnishings" organised by the Museum of Modern Art in New York. At this exhibition the elegant armchair made of hard rubber and fibreglass reinforced plastic, designed by the Eames husband and wife design team, also made an appearance. It was not amongst the prizewinners, but became famous under the name "La Chaise" and became the epitome of organic design of the period.[4] Here, as was the case with other famous examples, plastic was not only presented by recognised designers in one of the most renowned art museums in the world, it was also exhibited alongside contemporary art pieces by the likes of Henry Moore, Jean Arp and Gaston Lachaise, after whom the Eameses named this piece of furniture.

The "new" material was also greeted in East Germany shortly after the country was founded as a challenge to be embraced – both for cheap mass production and good design. The design and use of plastic was positively decreed by the state, with the backing of the renowned design schools in Halle, Burg Giebichenstein, Weimar and Berlin. The *plaste and elaste* articles as they became known, which were often designed with self-confident simplicity, flooded the country in the late fifties and early sixties.

Yet in West Germany many people found applying principles of design of plastic nothing short of painful. Hans Schwippert, who became the first president of the Deutscher Werkbund in 1953, put the crisis of conscience in a nutshell when he spoke of the "expulsion from the primal paradise" and of the "first major stage in the battle of the designers to find the well-shaped and respectable object,"[5] that was now over and on the verge of a new age. At the "Darmstädter Gespräch" in 1952 entitled "Man and Technology," groups of industrialists, philosophers, art-

Tupper-Gefäße in der Ausstellung »Neues Hausgerät in USA«, Stuttgart, 1951
Tupperware containers in the exhibition "New household articles in the USA,"
Stuttgart, 1951

Kunststoffteile für den Apparatebau: »Rein technische Teile (...) werden nicht form-
gestaltet und sind trotzdem oder deshalb meist schön«, Amtor Schwabe, 1953
Plastic components for appliances: "Purely technical components (...) are not
designed to be beautiful and most of them are nevertheless, or maybe for that
very reason, beautiful," Amtor Schwabe, 1953

neue Zeit einzutreten. Auf dem »Darmstädter Gespräch« von 1952 mit dem Titel »Mensch und Technik« suchte man sich im Kreise von Industriellen, Philosophen, Künstlern und »Formgebern«, vor allem Werkbundmitgliedern, der Verantwortung zu stellen, die mit neuen Technologien, verbesserten Herstellungsmethoden und nun nicht mehr zu verdrängenden Materialien entstand. Chancen und Bedrohung der modernen Technik wurden zu einem viel diskutierten Thema dieser Zeit, das auch die chemische Industrie mit einbezog. Der Kunsthistoriker Hans Sedlmayr weitete seine Bedenken sogar zu einer Generalabrechnung mit der Kunst seiner Zeit aus und diagnostizierte in seinem berühmten Buch *Verlust der Mitte* (1948) die Verwendung von Kunststoff als Krankheitssymptom der Menschheit: » (...) die Erkenntnis, dass der Schwerpunkt des Menschengeistes sich in die Sphäre des Anorganischen verrückt hat«, wofür » (...) das Erscheinen der künstlichen Werkstoffe ein großartiger Ausdruck ist«, zeige » (...) dass es sich (...) zweifellos (...) um eine Störung im Zustand des Menschen handelt.«[6]

Angesichts der Tatsache, dass die Glanzlichter der synthetischen Stoffe seit mehr als einem Jahrhundert existierten[7] und im Industriehistorismus und -jugendstil sowie im *Art déco* und amerikanischen *Styling* mit großen Auflagenzahlen Absatz fanden, erstaunt die Heftigkeit, mit der diese Diskussion zu einem so späten Zeitpunkt geführt wurde. Waren doch schon Mitte des 19. Jahrhunderts viele Kunststoffe durch chemische Umwandlung von organischen Naturstoffen wie Zellstoff, Kautschuk und Eiweiß gewonnen worden. Dadurch entstand eine Vielzahl von halbsynthetischen Materialien wie Weichgummi und der Hartgummi Ebonit, Vulkanfiber, Celluloid, Nitroseide, der Casein-Kunststoff »Galalith«, Kupfer- und Viskose-Kunstseide, Zellulose-Azetat, das Zellglas »Cellophan« und vieles andere mehr.

Bereits zu dieser Zeit und damit zu Beginn der Reformbewegungen im Kunstgewerbe wurde diese Diskussion auf ähnliche Weise schon einmal geführt. Von den Äußerungen zu den Urstoffen fanden die von Gottfried Semper in der Designgeschichte die größte Beachtung. In einer Periode des fieberhaften Experimentierens und Erfindens in der chemischen Industrie veröffentlichte Gottfried Semper, der viel zitierte Kommentator der ersten Weltausstellung 1851, die frühen Argumente für

ists and designers, and above all Werkbund members, had attempted to confront the responsibility that arose from new technologies, improved manufacturing methods, and materials that could no longer be ignored. The opportunities and threats posed by modern technology became a much-discussed general topic of the time, and the chemicals industry was part of the discussion. Art historian Hans Sedlmayr even extended the scope of his reservations, making a general reckoning with art of the time. In his famous book *Verlust der Mitte* (1948), he diagnosed the use of plastic as a symptom of mankind's sickness: " (...) the realisation that the focus of the human mind has shifted to the sphere of the inorganic [for which] (...) the appearance of synthetic materials is a magnificent expression (...) [shows] (...) that this is without a doubt (...) a disorder in the condition of human beings."[6]

In view of the fact that the celebrity synthetic materials had already been in existence for over a century[7] and were enjoying booming sales in styles ranging from industrial historicism, industrial *Jugendstil* to Art Déco and American *styling*, the vehemence with which these discussions were conducted at such a late date was astonishing. After all, back in the mid-19th century many synthetic materials had been made by chemically altering natural organic materials such as cellulose, rubber and protein. In this way, a great number of semi-synthetic materials were created, including soft rubber and the hard rubber ebonite, vulcanised fibre, celluloid, Chardonnet silk, the casein-based plastic Galalith, cuprammonium and viscose rayon, cellulose acetate, cellophane, to mention just a few.

There had been similar discussions at that time, at the beginning of the reform movements in arts and crafts. Of the comments made about primary materials, those by Gottfried Semper attracted most attention in the history of design. In a period of fervent experimentation and invention in the chemicals industry, Gottfried Semper, the often-quoted commentator on the first World's Fair in 1851, published early arguments in favour of rejecting synthetic materials. Semper was fixated on primary materials and their aesthetics, which is why he rejected new materials and new industrial methods. He saw modern industry's achievements as a threat: " (...) the hardest Porphyry and granite can be cut like chalk, polished like wax; ivory is softened and pressed into moulds, caoutchouc

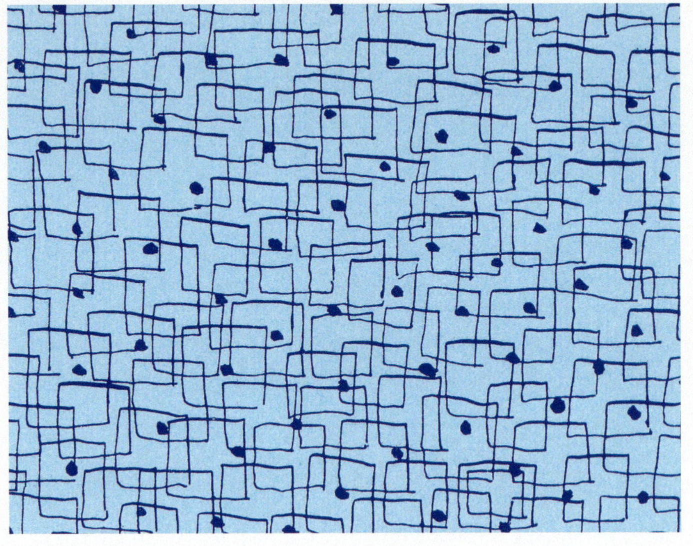

Jupp Ernst: Dekor-Entwürfe für Resopalplatten, vor 1955
Jupp Ernst: Designs for Resopal laminates, pre-1955

die Ablehnung der synthetischen Stoffe. Semper war auf Urstoffe und deren Ästhetik fixiert, weshalb er neue Materialien und neue industrielle Techniken ablehnte. Er betrachtete die modernen Leistungen der Industrie als Gefahr: »(...)der härteste Porphyr und Granit schneidet sich wie Kreide, poliert sich wie Wachs, das Elfenbein wird weich gemacht und in Formen gedrückt, Kautschuk und Guttapercha wird vulkanisiert und zu täuschenden Nachahmungen der Schnitzwerke in Holz, Metall und Stein genutzt, bei denen der natürliche Bereich der fingierten Stoffe weit überschritten wird.«[8] Ebenso wenig wie die aus der technischen Bewältigung des Materials resultierende Beliebigkeit der industriellen Form war für Semper die aus den synthetischen Materialien auf Kautschukbasis, dem »Faktotum der Industrie«, resultierende Form fassbar. Semper kapitulierte: »Bei einer solchen Materie steht einem Stilisten der Verstand still!«[9]

Neben der Gestaltung von Naturstoffen hat es schon früh auch deren Umformung gegeben. Das älteste Mittel, Naturstoffe in eine neue Substanz zu verwandeln, war das Feuer, das der Gewinnung und Bearbeitung von Metallen und Glas diente, der Umwandlung des weichen Tons in den keramischen Scherben. Dazu gehörte die gebrannte und gesinterte Erde wie Ton, Ziegel und Porzellan, der Stuck und Kalkbeton, aber auch Metall-Legierungen wie Bronze. Diese Fähigkeit, die den Menschen beflügelte und in der er sich als *homo faber* sah, rief zugleich aber auch immer den Gestalter auf den Plan, der in Ehrfurcht vor der Natur handeln wollte. Die Verwendung von natürlichen Materialien und der sie leitenden Formgebung wurden zum ideologischen Fundament von Streitpositionen divergierender Richtungen. Als ein Beispiel sollen die unterschiedlichen Designpositionen von John Ruskin und William Morris als Vertreter von *Arts and Crafts* und von Christopher Dresser als Vertreter einer zum *Aesthetic Movement* tendierenden Richtung angeführt werden.

Ruskin und Morris gingen noch von einer Naturauffassung im Sinne einer »natürlichen Natürlichkeit« aus, wie sie der Romantik entsprach. Morris war fasziniert von den Qualitäten des ungezügelten Wachstums, von Unregelmäßigkeit und freier Entfaltung, und begriff Natur als ein letztlich nicht erklärbares, vom menschlichen Willen unabhängiges und verehrungswürdiges Phänomen. Seine naturalistische Auffassung von Natur

and guttapercha are vulcanised and used for deceptive imitations of carvings in wood, metal and stone, in which the natural scope of the materials faked is vastly exceeded"[8]. Semper found the form that resulted from rubber-based synthetic materials, the "factotum of industry," just as incomprehensible as the arbitrariness of industrial form that resulted from technically mastering a material. Semper was forced to capitulate: "In the face of something like this a stylist's understanding is paralysed!"[9]

From early times, natural materials had not only been shaped but also transformed. The oldest means of transforming natural materials into new substances was fire, which was used both to make and process metals and glass, and to turn soft clay into hard ceramics. That included firing and sintering earth to make fireclay, tile or porcelain, stucco and lime concrete, but also metal alloys such as bronze – a skill that inspired mankind and in which they saw themselves as *homo faber*, but one which also always brought the designer, who was at pains to act with a sense of awe before nature, into the arena. The use of natural materials and an attitude to design that was guided by them became the ideological foundation of the conflicting positions adopted by divergent movements. An example of this can be seen in the different design positions of John Ruskin and William Morris as the representatives of *Arts and Crafts* and Christopher Dresser as the representative of a school of thought with leanings towards the *Aesthetic Movement*.

Ruskin and Morris still saw nature in terms of "natural naturalness," in the way the Romantic philosophers envisaged it. Morris was fascinated by the qualities of unbridled growth, by irregularity and free development, and saw nature as an ultimately inexplicable phenomenon that was independent of human will and worthy of being revered. His naturalistic view of nature was particularly evident in his two-dimensional decorative work on wallpaper and textiles, but also in his rejection of artificial dyes for colour printing and for dyeing textiles, which were becoming extremely popular. It is well known that he went to considerable lengths, in terms of time and cost, to have dyes delivered from beyond Europe, because plant-based dyes were scarcely available any more in his own country. Ruskin placed greater emphasis on the choice

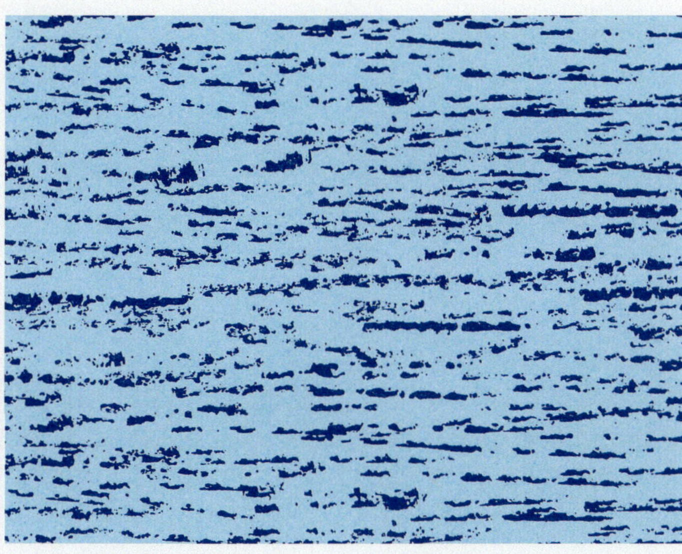

wurde besonders in der Ornamentik der Flächenarbeiten auf Tapeten und Textilien deutlich, aber auch in der Ablehnung der in seiner Zeit stark aufkommenden künstlichen Kolorierungen für den Farbdruck und die Färbung der Textilien. Es ist bekannt, dass er sich mit erheblichem Zeitaufwand und unter hohen Kosten Farben aus außereuropäischen Ländern anliefern ließ, weil Pflanzenfarben in seinem eigenen Land kaum mehr vorhanden waren. Bei Ruskin lag der Akzent stärker auf der Wahl traditioneller natürlicher Materialien für Gebrauchsobjekte, weil sie einer Werkästhetik entgegenkamen, in der sich der *living touch* der menschlichen Hand entäußerte. Der Kernsatz in *Seven Lamps of Architecture* lautet: »Ich glaube, die einzig entscheidende Frage bei allem Ornament ist einfach diese: War es mit Vergnügen und Genuss gemacht? War der Bildner glücklich, als er daran meißelte?«[10] Der Würde des Menschen entsprachen dann auch die besten Materialien, zumindest Materialien, die seine Handarbeit erkennen ließen, denn: »Alle gute Arbeit muss freie Handarbeit sein«.[11] So hatte Marmor für ihn einen minderen Wert, wenn er mit der Maschine geschnitten war. Seine Wertschätzung von Handarbeit ließ ihn erst recht Materialien verwerfen, die wie die halbsynthetischen Plastiken gegossen, geprägt und gestanzt waren.[12] Folglich lautete sein als Lehrspruch formuliertes Urteil: »Gusseiserne Ornamente sind barbarisch.«[13] Gusseisen war für ihn ein rohes und billiges Ersatzmittel von gehämmerter Schmiedearbeit, an der die Arbeit des Handwerkers zu erkennen war. Auf Ruskin geht die Regel von der Ehrlichkeit der Arbeit zurück, wozu auch die Ablehnung »falscher« Materialien gehörte wie die Bemalung von Holz, um Marmor vorzutäuschen.

Chemische Stoffe und ihre Herstellungsverfahren des Prägens, Stanzens und Gießens wurden besonders abgelehnt. Darüber hinaus traf sie von vornherein ein Verdikt, das die Grundprämisse der Ehrlichkeit der Produktkultur erschütterte. Sie galten als künstlich, und künstlich wurde mit falsch, mit trügerischem Schein, gleichgesetzt. Diese Scheinhaftigkeit, von den Gegnern als »Surrogat- und Talmikultur« bezeichnet, wurde besonders der viktorianischen Kunstindustrie, in Deutschland dem Gründerzeitstil zugeschrieben, da sich das Repräsentationsbedürfnis des ökonomisch schnell prosperierenden Bürgertums Status durch Nachahmung zu verschaffen suchte. Die Fabrikindustrie dieser Zeit ermöglichte

of traditional natural materials for articles for everyday use because they fitted in with an arts-and-crafts aesthetic that reflected the "living touch" of the human hand. The key sentence in *Seven Lamps of Architecture* is: "I believe the right question to ask, respecting all ornament, is simply this: Was it done with enjoyment – was the carver happy while he was about it?"[10] Thus the best materials were those that corresponded to man's dignity or at least allowed his craft to be recognised, since he believed: "All good work must be free hand-work."[11] Thus, marble had less value if it had been cut by machine. His estimation of work done by hand clearly led him to reject materials such as semi-synthetic materials that had been cast, embossed or punched.[12] Consequently the judgement he worded as an aphorism was: "Cast iron ornamentation is barbarous."[13] To him, cast iron was a cheap, coarse substitute for forged ironwork, which revealed the hand of the craftsman. The principle of faithfulness to the craft process goes back to Ruskin and also includes rejection of "false" materials such as painted wood intended to look like marble.

Materials made from chemicals and the processes used to produce them, such as embossing, punching and casting, were particularly vehemently rejected. They also passed a verdict on them from the very outset, claiming that they cast doubt on the basic premise of an honest product culture. They were considered to be artificial and artificial equated to a false, deceptive appearance. The fault for this falseness was laid at the feet of the Victorian art industry or in Germany the *Gründerzeit* style, which pandered to the need of the middle classes to display their rapidly burgeoning prosperity and to acquire status through imitation – labelled by their opponents a "culture of surrogate and sham." The mechanised industrial processes of this era already made it possible to produce higher volumes of goods, and factories also began to adopt these production methods, working with templates and casting moulds. Their products not only took on the appearance of the articles they were imitating but also the range of forms in the individual styles, usually with extravagant ornamentation.

The arts and crafts aesthetics of Christopher Dresser, a designer long ignored by design history, also required the work process to be visible.

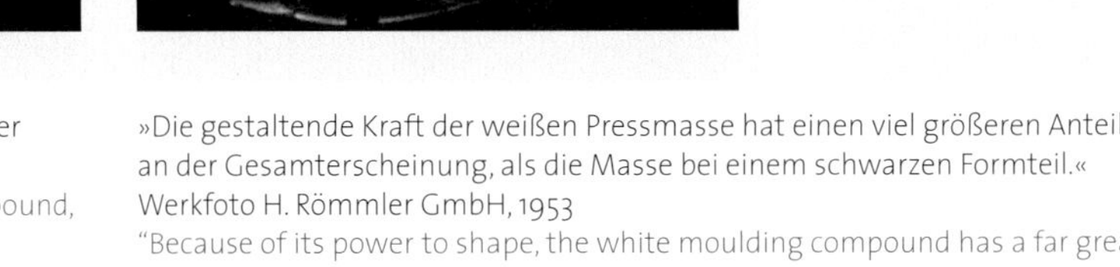

Doppelform für Telefonhörer aus Resopal-Pressmasse, Werkfoto H. Römmler
GmbH, 1953
Double mould for telephone receivers made from Resopal moulding compound,
company photo H. Römmler GmbH, 1953

»Die gestaltende Kraft der weißen Pressmasse hat einen viel größeren Anteil
an der Gesamterscheinung, als die Masse bei einem schwarzen Formteil.«
Werkfoto H. Römmler GmbH, 1953
"Because of its power to shape, the white moulding compound has a far greater
effect on the overall appearance than the compound of a black moulded article."
Company photo H. Römmler GmbH, 1953

bereits Auflagen in großer Stückzahl. Aber auch Manufakturen schlossen
sich dieser Konfektionierung an und arbeiteten mit Schablonen sowie
Press- und Gussformen. Ihre Produkte übernahmen nicht nur das Aussehen der Imitate, sondern auch das Formenspektrum der einzelnen Sti-
le, meist mit einem überbordenden Ornament.

In der Werkästhetik von Christopher Dresser, der lange von der Designgeschichte ignoriert worden war, wurde die sichtbare Arbeit des Produzenten nicht minder gefordert. Und doch agierte er in größerem Einklang
mit den industriellen Fertigungsmethoden und dem wissenschaftlichen
Geist seiner Zeit, dem er mit eigenen botanischen Studien auch selber
folgte. In den Gesetzen der Natur erkannte er die regulierende Ordnung,
die der Mensch durch seine Wissenschaft erschließen, nachahmen, in
manchen Fällen auch übertreffen konnte. Natur war durch Ordnung und
Gesetzmäßigkeit gekennzeichnet und deshalb auch ein verstandesmä-
ßig, das heißt wissenschaftlich erfassbares Terrain. Infolgedessen verlor
die wilde Natur aus wissenschaftlicher Sicht die Rolle des Vermittlers
von religiösen, moralischen oder sozialen Inhalten. Beiden Naturauffas-
sungen lagen unterschiedliche Orientierungen zugrunde: die regelhaf-
te und logische auf der wissenschaftlichen Seite Dressers, die freiheitli-
che und ursprüngliche auf der vitalistischen und naturalistischen Seite
Ruskins und Morris'.

Das Nachvollziehen natürlicher Formen besaß für Dresser keine geistige Kraft, dies war eben nicht die durch mathematische Gesetzmäßigkei-
ten, Symmetrie und Geometrie verbesserte und geordnete Natur.[14] Der
menschliche Geist erst war für ihn Maßstab aller Dinge. Daraus resultier-
te seine Vorliebe, immer neue ingeniöse Formen zu kreieren, um deren
über die bloße Abbildung hinausgehende Eigenständigkeit zu betonen.
Da die geistige Kraft des Menschen im Zentrum von Dressers Argumenta-
tion stand, kam es zur Wahl von Materialien und Formen, die den menschlichen Schöpferwillen im Sinne einer »Unterwerfung« von Natur deutlich
machten. So nutzte er das wegen seiner Kälte, Leblosigkeit und belie-
bigen Formbarkeit auch von den Handwerkergilden rigoros abgelehn-
te Gusseisen für seine Möbelentwürfe, oder auch Tafelsilber, ein mittels
Galvanisierungsverfahren hauchdünn mit Silber überzogenes Metall.
Dies wählte er, weil es edel erscheinende und doch preiswerte Objek-

And yet his actions were in fact in greater harmony with industrial manufacturing methods and with the scientific spirit of his time, which he
had tested out in a number of botanical studies. He saw in the laws of
nature a regulating order which mankind through his science was able
to access, imitate and, in some cases, even surpass. Nature was characterised by order and regularity and was therefore a domain that could
be comprehended by rationality, i.e. scientifically. Consequently, from
the scientific point of view, untamed nature lost its role as an example
to be followed in religious, moral and social matters. These two views of
nature were underpinned by different orientations: one based on order,
regularity and logic on the side of Dresser's scientific view, the other on
freedom and naturalness on the side of Morris and Ruskin's vitalist and
naturalistic view.

In Dresser's philosophy, the reproduction of natural forms did not pos-
sess any intellectual power, since it did not represent nature as being
improved and ordered by mathematical laws, symmetry and geome-
try.[14] For him it was the human mind that was the measure of all things.
From that ensued his predilection for creating new and ingenious forms
in order to emphasise his autonomy that went beyond mere depiction.
Since the artist's intellectual power was central to Dresser's argumen-
tation, he chose materials and forms that revealed the human desire to
create, in the sense of "subjugating" nature. So, the successful designer
used cast iron, a material that was also rigorously rejected by the crafts
guilds as being cold, lifeless and infinitely malleable. He chose silver plate
metal covered with a thin film of silver using a galvanisation process –
because it produced objects that looked precious but were inexpensive.
By contrast, the reform workshops of the *Arts and Crafts* movement did
not accept any kind of imitation. Since Dresser essentially accepted the
scientific achievements of his age, he expressly also assigned the chemicals industry a role in the creative process.

Much as the positions within the reform movements varied, there was
nevertheless a common ethos linking virtually all the different group-
ings, a belief that style should match the material. It was reflected in
their almost consistent philosophy of faithfulness to the material and
the craft process.

Jupp Ernst: Dekor »Filigran«, 1957 in die Kollektion der H. Römmler GmbH aufgenommen
Jupp Ernst: "Filigree" design that was taken into H. Römmler GmbH's collection in 1957

te gewährleistete und befand sich damit im direkten Gegensatz zu den Reformwerkstätten des *Arts and Crafts*, für die jegliche Nachbildung inakzeptabel war. Indem Dresser also die wissenschaftlichen Errungenschaften seines Zeitalters grundsätzlich akzeptierte, billigte er nachdrücklich auch der chemischen Industrie eine Rolle im Schöpfungsakt zu.

So unterschiedlich die Positionen innerhalb der Reformbewegungen auch waren, ein gemeinsames Ethos, dass der Stil dem Material entsprechen müsse, verband fast alle Richtungen. Dies schlug sich in der fast durchgängigen Orientierung an einer Material- und Werkgerechtigkeit nieder.

Mit der Inakzeptanz eines imitierenden Ersatzstoffes stimmte auch der liberale Christopher Dresser überein: »Der Versuch, einem Material das Aussehen eines anderen zu geben, ist eine Täuschung. Eine ‚Marmorierung' ist eine Täuschung; ein Bodenbelag, der einen Teppich oder eine Matte imitiert, ist eine Täuschung; ein Brüsseler Teppich, der einen türkischen Teppich imitiert, ist eine Täuschung; das gleiche gilt für einen Krug, der Riedgeflecht imitiert, für einen gedruckten Stoff, der einen gewebten imitiert, für eine Gaslampe, die sich als Öllampe präsentiert. Das sind alles Unwahrheiten im Ausdruck und sind, nebenbei gesagt, vulgäre Absurditäten, die umso bedauerlicher sind, als Imitationen stets weniger schön sind, als der eigentliche Gegenstand.«[15]

So stand also seit jeher die Wahl und Verwendung von bestimmten Materialien im Zentrum vieler Designdiskurse, auch im 20. Jahrhundert: Die Verwendung von Materialien orientierte sich in hohem Maße an den – pauschal so bezeichneten – Funktionalismus-Programmatiken, an den ethischen Prämissen der Kunstgewerbe-Reformbewegungen und der modernen Architektur, die in der zweiten Hälfte des 19. Jahrhunderts, zum Zeitpunkt lebhaften Erfinder- und Unternehmergeistes einsetzte und sich bis in die zweite Hälfte des 20. Jahrhunderts Geltung verschaffen konnte. Der Funktionalismus war und ist ein wertender Begriff, der keineswegs auf das meist verkürzt interpretierte Diktum von Louis Sullivan »form follows function«, wonach die Form lediglich der Funktion folgen soll, reduziert werden kann. Er war vielmehr eingebettet in eine Skala von ethischen Leitzielen, die die Gestaltung berücksichtigen musste, um zu einem guten Ergebnis zu kommen. Durch Designdiskurse

Even the liberally minded Christopher Dresser agreed that a material that imitated another was unacceptable: "All graining of wood is false, inasmuch as it attempts to deceive; the effort being made at causing one material to look like another which it is not. All "marbling" is false also: a floor-cloth made in imitation of carpet or matting is false; a Brussels carpet that imitates a Turkish carpet is false; so is a jug that imitates wicker-work, a printed fabric that imitates one that is woven, a gas-lamp that imitates an oil-lamp. These are all untruths in expression, and are, besides, vulgar absurdities which are the more lamentable, as the imitation is always less beautiful than the object imitated."[15]

Thus, the choice of particular materials has always been at the centre of much discussion on design, including during the 20th century. The use of materials was often connected to what goes under the blanket term of functionalism, to the ethical premises of the reform movements in arts and crafts and modern architecture, which began in the second half of the 19th century, when a spirit of lively invention and entrepreneurism prevailed, and by the second half of the 20th century had acquired recognition for itself. Functionalism was and is a term that carries a value judgement; it absolutely cannot be reduced to Louis Sullivan's dictate that "form follows function," which is usually interpreted simplistically. It was embedded in a scale of ethical objectives that design had to take into consideration in order to achieve a good result. Through design discourses and exhibitions, in other words through the world of people involved in design, a value system of "product morals" values was agreed that sometimes had a missionary and dogmatic character. Any divergence from the agreed doctrine would frequently be denounced as heresy.[16] With the concept of "good design" in the 1950s and 1960s, the practice of creating normative canons for designers, manufacturers and consumers was revived.

In the early 1950s, there was great interest in plastics in the newly founded West Germany. The reviewer of the Dusseldorf plastics show in October 1952 thus entitled his article that appeared in the Deutscher Werkbund's official magazine, *werk und zeit,* "The great opportunity." The show, at which 245 companies displayed their wares on 18,000 square metres and which attracted 165,000 visitors, enjoyed great popularity with the public. People were proud that Germany was gradually regaining the

Jupp Ernst: Resopalstand auf der Kunststoffmesse Düsseldorf, 1952
Jupp Ernst: Resopal stand at the Düsseldorf plastics show, 1952

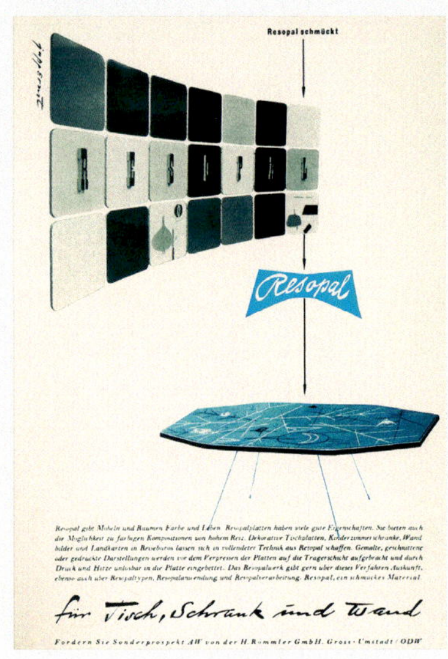

Jupp Ernst: Anzeige mit Motiven der Düsseldorfer Kunststoffmesse, 1952
Jupp Ernst: Advertisement with motifs taken from the Düsseldorf plastics show, 1952

und durch Ausstellungen, das heißt durch die Designöffentlichkeit, wurden »produktmoralische« Werte vereinbart, die bisweilen einen missionarischen und auch dogmatischen Charakter hatten. Ein Abfall von den Glaubenssätzen dieser Vereinbarungen wurde nicht selten als Häresie gebrandmarkt.[16] Mit dem Konzept der »Guten Form« in den fünfziger und sechziger Jahren des 20. Jahrhunderts lebte die Haltung, einen normativen Kanon für Designer, Produzenten und Konsumenten zu schaffen, noch einmal auf.

Anfang der fünfziger Jahre war das Interesse an Kunststoff in der jungen BRD groß. Als »die große Chance« titulierte daher der Rezensent der Düsseldorfer Kunststoffmesse im Oktober 1952 seinen Artikel im offiziellen Veröffentlichungsorgan des Deutschen Werkbundes, der Zeitschrift »werk und zeit«. Die Ausstellung, auf der 245 Unternehmen ihre Produkte auf 18.000 Quadratmetern zeigten und die 165.000 Besucher anzog, erfreute sich eines starken öffentlichen Interesses. Man war stolz, dass Deutschland im Begriff war, seine führende Stellung als Kunststofferzeuger in aller Welt zurück zu gewinnen, die es zu Anfang der dreißiger Jahr gehabt hatte. Mit Vehemenz versuchten die Aussteller, gegen den Eindruck zu argumentieren, Kunststoff sei ein minderwertiger, billiger Ersatzstoff. Daher wurde hier anstelle von Kunststoff der amerikanische Begriff »Plastics« oder der deutsche »Plastikmaterial« eingeführt, um falsche Assoziationen zu vermeiden.

Auch die *Deutsche Warenkunde* des Rates für Formgebung, herausgegeben von Mia Seeger und Stephan Hirzel unter der Schriftleitung von Heinz Löffelhardt, widmete ihren dritten Band, nach Keramik und Glas, dem Kunststoff. Parallel dazu organisierte der Deutsche Werkbund eine Reihe von Ausstellungen zur Materialkunde wie »Werkstoff Kunststoff« 1955 in Berlin.

Viele Anregungen, die materialtechnischen Hintergründe für die Öffentlichkeit darzustellen, gingen dabei von der kunststoffverarbeitenden Industrie selbst aus. Dr. phil. Gustav B. von Hartmann, der beratende Chemiker und wissenschaftliche Mitarbeiter der Resopalwerke H. Römmler GmbH in Groß-Umstadt, schrieb 1959 dem Deutschen Werkbund e.V.: »Ich weiß nicht, ob es in Ihren Rahmen passt, aber mir schwebt schon lange vor, einmal eine Ausstellung zu machen mit der Thematik ,Material und

leading position as a manufacturer of plastics that it had occupied in the early 1930s. There were emphatic attempts to argue against the impression that plastic was a downmarket, cheap substitute material and for that reason it was suggested that the commonly used German word *Kunststoff*, which translates literally as "synthetic material," be replaced by the English word "plastic" or that the English be modified to make a German word such as "Plastikmaterial," in order to avoid evoking the wrong kind of associations.

The *Deutsche Warenkunde,* a publication of the Rat für Formgebung (Design Council), edited by Heinz Löffelhardt with Mia Seeger and Stephan Hirzel, also dedicated its third volume to plastics, the first two having dealt with ceramics and glass. At the same time the Deutscher Werkbund organised a series of exhibitions on materials. One of them, entitled "Werkstoff Kunststoff", was held in Berlin in 1955 and was completely dedicated to plastics.

Many of the activities in this sphere aimed at giving the public background information on the material were initiated by the plastics processing industry itself. In 1959, Dr. Gustav B. von Hartmann, consultant chemist at the Resopalwerke H. Römmler GmbH in Gross-Umstadt/Odenwald, wrote to the Deutscher Werkbund e.V., Berlin: "I do not know if it fits in with your remit, but for a long time I have been toying with the idea of putting together an exhibition on 'material and design', in which the starting material – in the case of plastics usually completely unknown – (I do not mean the raw materials as such, but the so-called raw materials of plastic, in other words the materials that are processed into plastics) would be exhibited along with a number of key details about how they are processed and a small number of examples of ways of processing that are appropriate for this material and others that are not. It would not need to be exhaustive, and probably could not be, but it would be possible to pick out the most important and most characteristic plastics. This kind of topic suits the Werkbund to a tee, and would encourage endeavours to achieve good quality plastic products and thus be good advertising for plastics."[17] It is interesting to note that even in the late 1950s, the main objective of this kind of exhibition was to inform the public about plastics.

Pressartikel der H. Römmler GmbH,
Kunststoffmesse Düsseldorf, 1952
Moulded articles made by H. Römmler GmbH,
Düsseldorf plastics show, 1952

Form', in der also das Ausgangsmaterial, das ja meistens bei Kunststoffen völlig unbekannt ist, (also nicht die Rohstoffe, sondern die sogenannten Kunststoff-Rohstoffe, die Materialien also, die zu Kunststoffen verarbeitet werden) mit einigen charakteristischen Angaben über die Verarbeitungsweise und einigen wenigen Beispielen für materialgerechte gute und material-ungerechte Verarbeitung gezeigt wird; und das braucht und dürfte wahrscheinlich gar nicht vollständig sein, aber man könnte die wichtigsten und charakteristischsten Kunststoffmaterialien herausgreifen. Eine solche Thematik wäre ja doch sozusagen dem Werkbund auf den Leib geschrieben, und würde die Bemühung um die Qualität in Kunststofferzeugnissen fördern und damit der Werbung für Kunststoffe dienen.«[17] Noch Ende der fünfziger Jahre hatte eine solche Ausstellung also primär die Aufgabe, über das synthetische Material aufzuklären.

Doch die Grundsatzdebatten betrafen nicht nur die Verwendung, auch der Kunststoff an sich war Gegenstand heftiger Dispute. Hans Schwippert sprach von der »Charakterlosigkeit« des neuen Stoffes, der dem Gestalter oft nicht den geringsten Widerstand entgegensetze. 1953 betitelte er seinen Artikel in *Baukunst und Werkform* mit »Das Ende der Werkgerechtigkeit« und machte seinen Kollegen den Vorwurf: »Ihr habt vor lauter Verliebtheit in den Stoff das freie, spielerische Formen vergessen.« Ein »grundlegendes Prinzip des anständigen Bildens und Formens und Herstellens« seien »Werksicherungen« und »Materialgerechtigkeit«.[18]

Schon auf den Darmstädter Gesprächen von 1952 hatte Hans Schwippert konstatiert: »Was da an neuen Stoffen vor uns steht, ist in einem Maße willfährig uns gegenüber, wie wir das bisher nicht gekannt haben. Da gibt es plastische Stoffe, die sagen gar nicht mehr: Ich bin so, und du, Gestalter, du Hersteller musst sehen, wie du mit mir fertig wirst; sondern diese neuen Stoffe sagen: Bitte schön, ein Esslöffel von dem, ein Teelöffel von jenem, und ich bin wieder anders. Ich bin so, – ich bin so, – ich bin so! – Es ist deine Sache, das zu bestimmen. Die Stoffe bringen gar keine spezifischen, strengen Charaktere auf uns zu, sondern sie sagen: Bitte schön, du bist der Herr, ich bin der Diener, ich tue völlig, was du willst. Das hatten wir bisher nicht. Bisher konnten wir das Holz nur so weit biegen, und dann brach's. Aber diese Stoffe sagen: Bitte schön, gib mir eine Prise davon oder eine Messerspitze hiervon, und ich tue, was du willst.

However, the fundamental debate was not only about whether the use of plastic was legitimate; the material itself was also subject of vehement dispute. Hans Schwippert spoke of the "lack of character" of the new material that often did not even offer the designer the slightest resistance. In 1953, he called his article in *Baukunst und Werkform* "The end of faithfulness to the craft process" and reproached his colleagues as follows: "You are so in love with the material that you have completely forgotten about free and playful forms." He went on to remind them that: "faithfulness to the craft process" and "truth to material" are a "fundamental principle of proper form and design and production."[18]

As early as the Darmstädter Gespräche in 1952 Hans Schwippert had observed: "The new materials we are faced with now are submissive to an extent we have never before experienced. These plastic materials no longer say: I am like this, and you, Mr. Designer or Mr. Manufacturer, have to deal with me. No, these new materials say: A dessertspoon of this please, a teaspoon of that, and I will become something completely different. I am like this, – I am like this, – I am like this! – It is up to you to make the decision. The materials do not present us with any specific, strong characters; they simply say: All right, you are the master, I am the servant, I shall do exactly what you want. We have never had that situation before. In the past we were only able to bend the piece of wood to a certain point and then it would break. But these materials say: Give me a pinch of this or a dash of that and I will do whatever you like. We are faced with the responsibility of sovereignty over materials that we have never had before and thus with a vastness that we are not quite equipped to deal with."[19] Although the focus of the debate on plastic had now shifted to the question of the designer's greater responsibility, " (…) an infinite expansion of our sphere of power! This brings with it an infinite expansion of the designer's responsibility towards the material (…)"[20] as Schwippert remarked, the peculiar properties of the material that had to date proved difficult were nevertheless described point for point.

It was often said that it was the "lack of structure of plastics [that can] (…) create virtually any property, form or structure (…) [that] causes the problems. This versatility is (…) the greatest advantage but also the danger of plastics."[21] The complete malleability of the material endangered

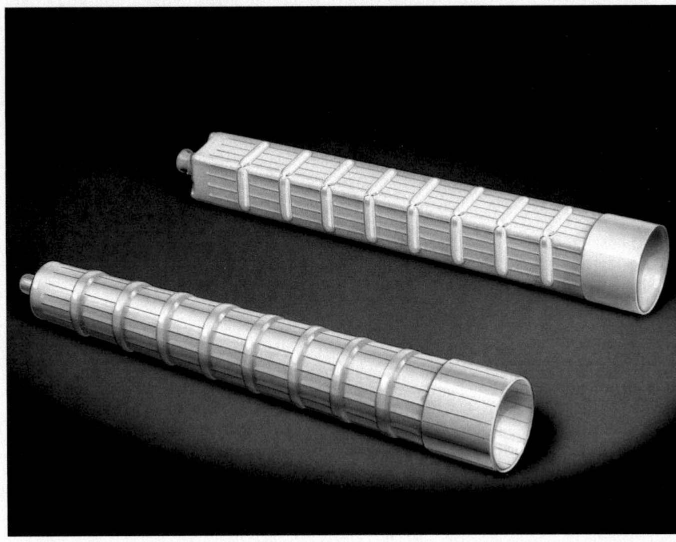

H. Römmler GmbH: Thermoplast-Spritzteile für einen Geschirrspülautomaten, sechziger Jahre
H. Römmler GmbH: Injection-moulded thermoplastic components for a dishwasher, 1960s

H. Römmler GmbH: Plattenspieler-Deckel aus Thermoplast, sechziger Jahre
H. Römmler GmbH: Injection-moulded thermoplastic cover for a record player, 1960s

Wir stehen vor einer Aufgabe der souveränen Beherrschung der Stoffe, die wir bisher nicht hatten, und sind damit vor einer Weite, für die wir uns nicht ganz gerüstet fühlen.«[19]

Obwohl die Auseinandersetzung mit Plastik nun vor allem die größere Verantwortung des Designers unterstrich – »(...) eine unendliche Vergrößerung unseres Machtbereichs! Damit eine unendliche Vergrößerung der Verantwortung des Gestalters der Stoffe (...)«[20] wie Schwippert betont – wurden auch die besonderen und bisher problematischen Eigenschaften des Stoffes dezidiert beschrieben.

Es war häufig die Rede davon, dass die »Strukturlosigkeit der Kunststoffe«, die sich »(...) fast jede beliebige Eigenschaft, Form und Struktur geben lassen (...)«, problematisch sei. »In dieser Vielseitigkeit (...)« liege »(...) der große Vorzug, aber auch die Gefahr der Kunststoffe«.[21] Die willkürliche Plastizität des Stoffes gefährdete in ihren Augen die Voraussetzungen der »Werkgerechtigkeit«, wonach die Hand des Gestalters sich vom Material leiten ließ. Die Protagonisten der frühen Reformbewegungen sahen diese Gefahr in dem Einsatz der Maschine, die sich dem Material nahezu willkürlich aufoktroyierte. Mit zunehmender Akzeptanz der Maschine verlagerte sich der Akzent der Forderung nach einer Werkgerechtigkeit auf die Übereinstimmung von Werkvorgang und Material.

Als Industrieprodukt *par excellence* verfügte Plastik über keine Werkstoff-Struktur, die eine individuelle Behandlung des Stoffs widerspiegelt. Im Gegenteil, es war gänzlich abhängig von der Maschine und hatte damit keinen Werkcharakter, der aus dem Zusammenspiel von menschlicher Hand und dem Widerständigen oder zumindest dem Charakter des Materials resultiert. Plastik galt als bindungsloser Stoff: »Er ist kein Werkstoff mehr im alten Sinn des Handwerks, ein Stoff, den die Hand werkend verändert und zum Erzeugnis formt. Man kann manche Erzeugnisse aus Kunststoff noch weiter *verarbeiten*, aber nicht *bearbeiten*. Form, Oberfläche und Farbe, diese drei Faktoren, die das Bild eines Gegenstandes ergeben, sind beim Erzeugnis aus Kunststoff nach der Eigenart seiner Herstellung endgültig fertig ausgebildet, wenn das Teil aus der Presse kommt.«[22]

Nur sehr selten differenzierte man in generellen Verlautbarungen über »industrielle Formgebung« bei den Kunststoffmaterialien. Eine Aus-

in their eyes the conditions underlying "faithfulness to the craft process" by which the hand of the designer was guided by the material. The protagonists of the early reform movements had considered this to be the danger of machines that imposed themselves almost arbitrarily on the material. As machines became more accepted, the focus of the demand for "faithfulness to the craft process" shifted to concurrence between the craft process and the material.

As an industrial product *par excellence*, plastic did not have the structure of a material that is produced by being individually worked. On the contrary, it was entirely dependent on the machine, which meant that it lost its crafted character, something produced by the interplay between the human hand and the resistance, or at least the character, of the material. Plastic was considered to be a non-relational material: "It is not a material in the old sense of a traditional craftsman, i.e. a material that is changed by the work of the human hand and shaped into an article. Some articles of plastic can still be further processed, but they cannot be worked. Form, surface and colour, these three factors dictate what an object looks like: but, by the time a plastic article comes out of the press, all those things have been determined once and for all by the way the article has been produced."[22]

Only rarely did general statements make the distinction of "industrial" design for plastics. Industrial designer Wilhelm Braun-Feldweg was an exception with his publication of 1954 entitled *Normen und Formen industrieller Produktion* (Standards and forms of industrial production). Here he discussed plastics under the title "The most obedient material,"[23] but also pointed out that some of the "new materials," such as celluloid and Plexiglas, are suited to "being crafted or machined." However, he went on to say, a great handicap were the extremely complex and expensive injection moulds that were used to make the cheap mass-produced articles. People had to therefore think very carefully about whether the investment was worthwhile. "For industrial mass production of our time the overly sophisticated tool is a major issue. This is where the risk is concentrated. Because forms that do not turn out to have popular appeal will incur heavy losses."[24] Nevertheless, his publication does present the advantages of the "plastic material" most impressively and illustrates the

Plexiglas-Behälter aus britischer Produktion, 1954 von Wilhelm Braun-Feldweg
veröffentlicht
British-made Plexiglas containers, published in 1954 by Wilhelm Braun-Feldweg

nahme stellt der Industriedesigner Wilhelm Braun-Feldweg mit seinem 1954 erschienen Werk *Normen und Formen industrieller Produktion* dar. Hier setzte er sich mit Kunststoff zwar auch unter dem Titel »Der fügsamste Werkstoff«[23] auseinander, wies aber darauf hin, dass einige der »neuen Werkstoffe« wie Zelluloid und Plexiglas zu »spanabhebender und handwerklicher Bearbeitung« geeignet seien. Ein großes Handicap seien allerdings die extrem aufwendigen und teuren Spritzgussformen, mit denen die billigen Massenartikel hergestellt werden. Es müsse deshalb genau überlegt werden, ob sich diese Investition lohnt. »Für die industrielle Massenproduktion unserer Zeit bedeutet das hochgezüchtete Werkzeug einen Brennpunkt. Hier konzentriert sich das Wagnis. Denn Formen, die keinen Anklang finden, verursachen schwere Verluste.«[24] Dennoch führt seine Publikation wohl am eindrucksvollsten die Vorteile des »Plastik-Materials« vor und belegt die Vorteile des neuen Stoffes durch Fotoabbildungen bester Gestaltungsbeispiele meist aus den USA, aber auch aus England und Schweden.

Das Verdikt der Strukturlosigkeit bezog sich auch auf die Oberflächenstruktur des Materials. Immer wieder wurde nach Möglichkeiten gesucht, die negativen Seiten des Kunststoffs durch entsprechende Bearbeitung auszugleichen. Aber der Eindruck von der Sterilität des Materials blieb. »Die ‚billige' Wirkung solcher Gegenstände hat nicht im Werkstoff ihre Ursache, sondern im mangelnden Verständnis der Hersteller für dessen Eigenart und besondere Möglichkeiten.«[25] Die Beliebigkeit seiner Formbarkeit erlaubte keine Zuordnung gewohnter Assoziationen. »Bezeichnenderweise werden deshalb gerade bei den ‚weichen' oder ‚plastischen' Arten am häufigsten organische Strukturen imitiert, um dem Gefühl eine gewisse Hilfestellung zu leisten. Ihre Richtungslosigkeit aber, die oftmals unangenehme Griffigkeit und der Nachteil, dass sie (als Folie) nicht atmen, lassen sich dadurch nicht beseitigen. Es haftet ihnen häufig etwas Provisorisches und Steriles an, das sich wohl mit einer Klinik, aber wenig mit einem Wohnraum verträgt.«[26]

Der Designer und Direktor der Werkkunstschulen in Wuppertal und Kassel, Jupp Ernst, wies zwar ebenfalls auf die negativen Seiten der Kunststoffoberflächen hin, die ihm zu glatt, glänzend und daher zu unsinnlich erschienen. Er stellte jedoch überzeugende amerikanische Beispiele

benefits of the new material with photographs of the best design examples, mostly from the USA, but also from Britain and Sweden.

The "lack of structure" verdict also applied to the surface of the material. Numerous attempts were made to find ways of compensating for the negative aspects of plastic by appropriate finishing. But the sterile effect of the material remained. "The "cheap" effect of these objects stems not only from the material but also from the manufacturer's lack of understanding of its peculiarity and specific possibilities."[25] Because the material could be made into any shape, he [the manufacturer] did not have the accustomed associations with the materials. "It is typically the "soft" or "plastic" types of plastic that most often imitate organic structures in order to help our feelings somewhat. However, their lack of direction, the fact that they are often unpleasant to touch, and the disadvantage that they (as a film) do not breath, cannot be eliminated in this way. They often have a provisional and sterile look about them, which is well suited to a clinic but not to a domestic setting."[26]

The designer and director of the schools of applied arts in Wuppertal and Cassel, Jupp Ernst, also pointed out the negative aspects of plastic surfaces, which he found too smooth and shiny and therefore unsensuous. He presented convincing American examples, where a manufacturing process had given the surface a perceptible texture. That created, he said, a certain aesthetic appeal. Ernst even advised recourse to mechanical processes to create a roughened, fluted or stippled effect, which would give the surface of the plastic a relief character, and thus enhance its haptic properties. He also believed it was legitimate to use contemporary ornament and was in favour of adding patterns to the surfaces. He designed many patterns for Resopal laminates in conjunction with his colleague Ernst Oberhoff.[27]

The fact that organic components of plastic do not acquire the characteristics of inorganic material – even after the manufacturing process is complete – can cause slight changes to the surface: "Since the reflective quality of smooth, even surfaces means that the slightest irregularity is instantly noticeable, these even surfaces should be avoided and a decorative line that may often be condemned as superfluous may serve a good purpose, just as appropriate ornament may well be valuable, par-

Ernst Oberhoff: Künstlerische Experimente von Aquarellmalerei unter Resopal, Anfang fünfziger Jahre
Ernst Oberhoff: Artistic experiments with watercolour paintings under Resopal surfaces, early 1950s

vor, bei denen der Herstellungsprozess den Oberflächen eine wahrnehmbare Struktur gab und damit einen ästhetischen Reiz. Ernst riet sogar zu Hilfsmitteln, die durch mechanische Aufrauung, Riffelung und Punktierung der Kunststoffoberfläche einen Reliefcharakter für stärkere haptische Eigenschaften geben konnten. Auch hielt er die Anwendung eines zeitgemäßen Ornaments für legitim und sprach sich für die Musterung der Flächen aus. Zusammen mit seinem Kollegen Ernst Oberhoff hat er viele Muster für Resopalplatten angefertigt.[27] Infolgedessen, dass die organischen Bestandteile des Kunststoffs auch nach dem Abschluss des Herstellungsprozesses nicht die Eigenschaften anorganischen Materials annehmen, entstehen häufig leichte Veränderungen an den Oberflächen: »Da nun bei ebenen, glatten Flächen infolge der Spiegelung die feinste Unebenheit sofort erkennbar ist, sollte man diese Ebenen vermeiden, und manche Zierlinie, die nicht selten als überflüssig verurteilt wird, kann ihren guten Sinn haben, wie überhaupt das sinnvolle Ornament gerade bei Kunststoffen durchaus wertvoll ist, weil eben dem Kunststoff jedes eigene Leben, wenn man so sagen darf, fehlt.«[28]

Eine ganz besondere Chance sah man in der Anwendung von Farben. Hans Schwippert bemerkte geradezu euphorisch: »Überhaupt – die Farben! So etwas wie ein Farbenrausch, eine neuerwachende Farbfreudigkeit kündigt sich an, inmitten dieser chemisch gereinigten Welt. Auch hier winken große Möglichkeiten, die nur erkannt, gebändigt, gestaltet werden wollen. Hier muss eine neue Verantwortlichkeit in Kraft treten.«[29]

Mit den sechziger Jahren kam es zum Durchbruch des künstlichen Materials. Vor allem die Gestaltungsformen in den USA hatten zu einer fast uneingeschränkten Akzeptanz des Kunststoffes geführt. Designer wandten das Material nicht mehr nur in Sonderbereichen an, sondern auch gängige Materialien wie Metall wurden nun durch Plastik ersetzt. Ganze Industriezweige, ja nationale Produktionsschwerpunkte entstanden. Technikeuphorie und soziale Utopien inspirierten die Designer bis Anfang der siebziger Jahre, mit Plastik zu arbeiten. Die Ölkrise 1973 führte dann zwar zu einem Wandel in der Haltung gegenüber Kunststoffen, aber deren Bedeutung für Alltagskultur und Design blieb groß.

Steine aus bemaltem Schaumstoff zu verwenden, um den Eindruck eines Kiesbeetes zu simulieren, das man in Form von Fliesen ins Bad legen

ticularly in the case of plastics, because plastics – if I may say so – have no life of their own."[28]

The use of colour was seen as a particular opportunity. Hans Schwippert remarked with a sense of euphoria even: "Oh, colours! Something akin to a frenzy of colours, a revived colourfulness, has been heralded in the midst of this chemically clean world. And here too great opportunities are beckoning: they merely have to be recognised, harnessed, and given shape. Here a new kind of responsibility must come into force."[29]

The 1960s saw a positive breakthrough for synthetic materials. In particular, design forms in the USA had brought about virtually unreserved acceptance of plastic. Designers no longer used the material exclusively in special areas; now common materials such as metal were also replaced by plastic. Entire industries grew up, with particular countries concentrating on particular specialties. Until the early seventies, euphoria over technology and social utopias inspired designers to work with plastic. Although the oil shock in 1973 brought about a universal change in the attitude to plastic, it was too well established by then to be toppled from its high-profile place in everyday culture and design.

Using stones made of foamed polyurethane, painted to simulate a bed of pebbles that could be laid as tiles in the bathroom, producing armchairs made of only wafer-thin transparent PVC, such as the Italian »Blow« chair, seating that did not take shape until you opened the packing box – all this and much more besides not only testified to how people enjoyed new things, but was also fed by a desire to be provocative. People were almost completely uninhibited in their experimentation with plastics. The brightly coloured plastic laminates, which the Memphis group used for furniture in the 1980s, were also deliberately chosen to re-evaluate bad taste. The idea was to once and for all shatter the rules of Modernism, its normative canon, and exhaust the potential of the new era, something which had been one of the ambitions of Classical Modernism.

The play on meaning of the oppositional designers who used plastic as a counterargument against Modernism resulted in it being used by all opponents of Modernism of every possible hue. While the "enemy" of early Modernism had been the *Gründerzeit* style in Germany and the foe of *Arts and Crafts* had been the Victorian art industry, Modernism in

konnte, Sessel zu produzieren, die nur aus einem Hauch von dünnem, transparentem PVC bestanden, wie der italienische »Blow«, Sitzgelegenheiten, die sich erst zu ihrer Form entfalteten, wenn man die Verpackungsschachtel öffnete – das alles und vieles mehr zeugte nicht nur von der Lust am Neuen, sondern auch von dem Wunsch nach Provokation. Nahezu hemmungslos wurde nun mit Kunststoff experimentiert. Auch die bunten Plastik-Laminate, die die Gruppe Memphis in den achtziger Jahren für Möbel geltend machte, wurden bewusst als Umwertung des schlechten Geschmacks gewählt. Die Regeln der langanhaltenden Moderne, der normative Kanon, sollte endgültig gesprengt und das Potential der neuen Zeit, ein Anspruch der historischen Moderne selbst, ausgeschöpft werden.

Die Gegnerschaft zur Moderne manifestierte sich also unter anderem in der Anerkennung der Verwendung von Kunststoff. Dies traf auf sämtliche Gegen- und Nebenpositionen der Moderne zu und erlaubte den Oppositionen ein facettenreiches Spiel mit unterschiedlichsten Bedeutungen. War das Feindbild der aufkommenden Moderne der Gründerzeitstil in Deutschland und von *Arts and Crafts* die viktorianische Kunstindustrie, so grenzte sich die Moderne in den zwanziger und frühen dreißiger Jahren vom *Art Déco* und amerikanischen *Styling* ab und in den späteren Jahren vom »Nierentischstil« der Fifties. Alle nicht oder wenig anerkannten Richtungen haben ihre Liberalität und Experimentierfreude an dem künstlichen Material erprobt, aber die strenge Moderne behielt durch ihren allzu orthodoxen »Willen zur Wahrheit«, wie Michel Foucault es formulierte, einen blinden Fleck und verspielte dadurch teilweise die Möglichkeiten, mit Hilfe von Kunststoff ihre eigenen Ansprüche einzulösen und günstige Massenprodukte herzustellen.

the 1920s and early 1930s was keen to distance itself from *Art Déco* and American *styling*, and in the 1950s from the *"Nierentisch"* style. Every movement that had been rejected or given little recognition tested out its liberality and joy in experimentation on the synthetic material. Strict Modernism had a blind spot caused by its all too orthodox "will for truth", as Michel Foucault put it, and to some extent wasted the opportunity to use plastic to achieve its own ambition and mass-produce inexpensive products.

Jupp Ernst: Markenzeichen Resopal, Reinzeichnung, 1951
Jupp Ernst: Resopal trade mark, finished artwork, 1951

DIE ERSCHAFFUNG DER MARKE RESOPAL DURCH JUPP ERNST
THE CREATION OF THE RESOPAL BRAND BY JUPP ERNST

Nach Kriegsende musste die H. Römmler AG nach Demontage und Verstaatlichung ihres ostdeutschen Stammsitzes, als GmbH in Groß-Umstadt wieder ganz neu anfangen. Künstlerischer Berater wurde hier 1952 Jupp Ernst (1905–1987), der nicht nur Werbe- und Gebrauchsgrafiker, sondern auch ein innovativer Produktdesigner, Werkbundmitglied, Direktor der Werkkunstschule Wuppertal und Mitinitiator des 1951 gegründeten Rats für Formgebung war. Ihm oblag die Präsentation der gesamten Produktpalette. Er entwarf das schwungvolle Resopal-Markenzeichen, das bis 1967 Gültigkeit hatte, vor allem aber Dekore, Prospekte, Anzeigen und Messeauftritte. Er entwickelte präzise Werbepläne, »genau hingestimmt auf den Empfängerkreis« und gestaltete 1953 einen Ausstellungsraum zur Präsentation der neuen Erzeugnisse.

Auch wenn Jupp Ernst Resopal zur berühmten Marke gemacht hat, war die Zusammenarbeit mit der H. Römmler GmbH für ihn keineswegs immer erfreulich. Die meisten seiner Vorschläge wurden nicht positiv aufgenommen, geschweige denn realisiert. So erging es zum Beispiel seinem Plan von 1953, einen Wettbewerb zur Herstellung von Möbeln aus Resopal unter führenden Architekten wie Hans Schwippert, Georg Leowald, Egon Eiermann und Hans Scharoun zu veranstalten. Die Begründung seiner Idee lautete: »Resopal wird bisher vornehmlich als Furnierplatte verwandt. Die Anwendung unterscheidet sich also nicht wesentlich von den bisher im Möbelbau üblichen Techniken. Im Resopal liegen aber neue technische Möglichkeiten, die zu entwickeln sind. Die Ausschreibung soll die stagnierende Entwicklung im Möbelbau in Fluß bringen und dem Resopal neue eigenständige Anwendungsmöglichkeiten erschließen.« Neben diesem Ziel hatte Jupp Ernst auch die Werbewirksamkeit eines solchen Wettbewerbs vor Augen. Die Ergebnisse der Ausschreibung wären für die Veröffentlichung in angesehenen Architekturzeitschriften ideal gewesen. Die Möbel hätten in wichtigen Ausstellungen wie zum Beispiel der Mailänder Triennale gezeigt werden können. Für die Herstellung der Resopal-Möbel schlug er die damals marktführenden Firmen Deutsche Werkstätten und WK-Möbel vor.

Schon 1953 wollte Jupp Ernst der H. Römmler GmbH die Zusammenarbeit aufkündigen, ließ sich dann aber überreden weiterzumachen, bis er sich 1955 endgültig zurückzog.

After the War, H. Römmler AG was first dismantled by the Soviet occupying forces and then nationalized by the East German government. It moved to Gross-Umstadt, near Frankfurt in West Germany and made a fresh start as a limited company. In 1952, Jupp Ernst (1905–1987) became the company's artistic consultant. He was not only a commercial and advertising artist, but also an innovative product designer, *Werkbund* member, director of the *Werkkunstschule*, a school of applied arts in Wuppertal, and co-initiator of the *Rat für Formgebung* which was founded in 1951. He was responsible for the presentation of the company's entire product range. He designed the dynamic Resopal trademark, which was valid until 1967, but primarily he designed the company's brochures, advertising, displays for trade show and the different laminate patterns. He developed precisely honed advertising campaigns "carefully tailored to the target group" and in 1953 designed a showroom for displaying new products.

However, although Jupp Ernst made Resopal a famous household brand, the collaboration with H. Römmler GmbH was not always a happy one for him. Most of his suggestions were not even viewed positively, let alone put into practice. That was the case, for example, in 1953 with his plan to hold a competition for furniture made of Resopal and invite leading architects such as Hans Schwippert, Georg Leowald, Egon Eiermann and Hans Scharoun to take part. His reasoning was as follows: "To date Resopal has been primarily used as a veneer. Thus the way it is used differs little from the usual methods of furniture making. However, Resopal offers new technical possibilities, which need to be developed. The idea behind the competition is to inject some impetus into the currently stagnating furniture industry and explore new ways of using Resopal that are not connected to traditional ones." Apart from this objective, Jupp Ernst also had the advertising effect of a competition of this kind in mind. The results would have been ideal for publication in well-respected architectural journals. It would have been possible to exhibit the furniture at important shows such as the Milan Triennale. He suggested that the Resopal furniture be manufactured by the two market leaders of the time: Deutsche Werkstätten and WK-Möbel.

Jupp Ernst wanted to terminate his collaboration with H. Römmler GmbH in 1953, but was persuaded to continue. He finally retired in 1955.

■ **JUPP ERNST: STAND DER H. RÖMMLER GMBH AUF DER KUNSTSTOFF-MESSE DÜSSELDORF, 1952**

Zum ersten Mal wurde hier einer breiten Öffentlichkeit die Verwendung von Resopal-Kunststoffplatten in größerem Umfang vorgeführt. Jupp Ernst präsentierte sie räumlich inszeniert vor einem neutralen Stoffhintergrund. Die rotierenden Schriftelemente zeigten im Wechsel die Worte Resopal und Römmler. Für den Außenbereich schuf Jupp Ernst eine über sieben Meter hohe Skulptur des Firmenzeichens. Die Messe stand unter dem Motto »Wunder der Kunststoffe«.

■ **JUPP ERNST: STAND FOR H. RÖMMLER GMBH AT THE DÜSSELDORF PLASTICS SHOW, 1952**

This was the first time that Resopal laminated panels used on a large scale were presented to the general public. Jupp Ernst created a three-dimensional display against a neutral fabric background. The rotating lettering alternately displayed the words Resopal and Römmler. For the outdoor area Jupp Ernst created a sculpture of the company logo that stood over seven metres tall. The show's motto was "The miracle of plastic."

■ **JUPP ERNST: KUNSTSTOFFMESSE DÜSSELDORF 1952 UND 1955**

In der Entwurfszeichnung für den Stand von 1952 sind die unterschiedlichen Schauelemente und der ebenfalls von Jupp Ernst entworfene große Tisch mit schwarzer Resopalplatte in einer heiter-lockeren Atmosphäre dargestellt. Das Foto zeigt den Messestand der H. Römmler GmbH von 1955. Im Regal wurde auch ein von Jupp Ernst entworfenes Frühstücksgeschirr aus Resopalpressmasse präsentiert, das in sieben verschiedenen Farben gehalten war.

■ **JUPP ERNST: THE DÜSSELDORF PLASTICS SHOW, 1952 AND 1955**

In the design drawing for the 1952 stand the different display elements and the large table with a black Resopal top, also designed by Jupp Ernst, are shown in a lively and relaxed setting. The photo shows H. Römmler GmbH's stand at the 1955 show. A set of breakfast tableware made from moulded Resopal in seven different colours, which Jupp Ernst had also designed, was displayed on the shelving.

Der Wunsch nach einer heiter abgestimmten Farbigkeit unserer Räume kann in vollendeter Weise erfüllt werden. Die farbigen Resopalplatten haben als Furnierplatten für den Möbelbau und als Wandverkleidungsplatten hervorragende Eigenschaften: Resopalplatten sind poren- und rißfrei, kratzfest, lichtecht und abwaschbar. Fordern Sie bitte Sonderprospekt BIV an, der Sie über Eigenschaften, Verwendungs- und Verarbeitungsmöglichkeiten unterrichtet.
H. RÖMMLER GMBH GROSS-UMSTADT ODENWALD

Resopal
für Tisch, Schrank und Wand

Resopal
gefällt mir
Eine Oberfläche, glänzend wie japanischer Lack oder matt wie Seide, dazu eine reiche Skala schöner Farben! Resopal gefällt auf den ersten Blick. Als ein hochgezüchtetes Produkt der Kunststoffchemie ist Resopal natürlichen Werkstoffen überlegen. Es vereinigt in sich viele Eigenschaften: Resopal ist abwaschbar und lichtecht, schlag- und kratzfest. Eine Sonderqualität ist selbst gegen Zigarettenglut unempfindlich. Resopal ist als Fertigplatte in sich schön und vollendet und bedarf keiner weiteren Nachbehandlung.
H. RÖMMLER GMBH · RESOPALWERKE · GROSS-UMSTADT
Resopal
für Tisch, Schrank und Wand

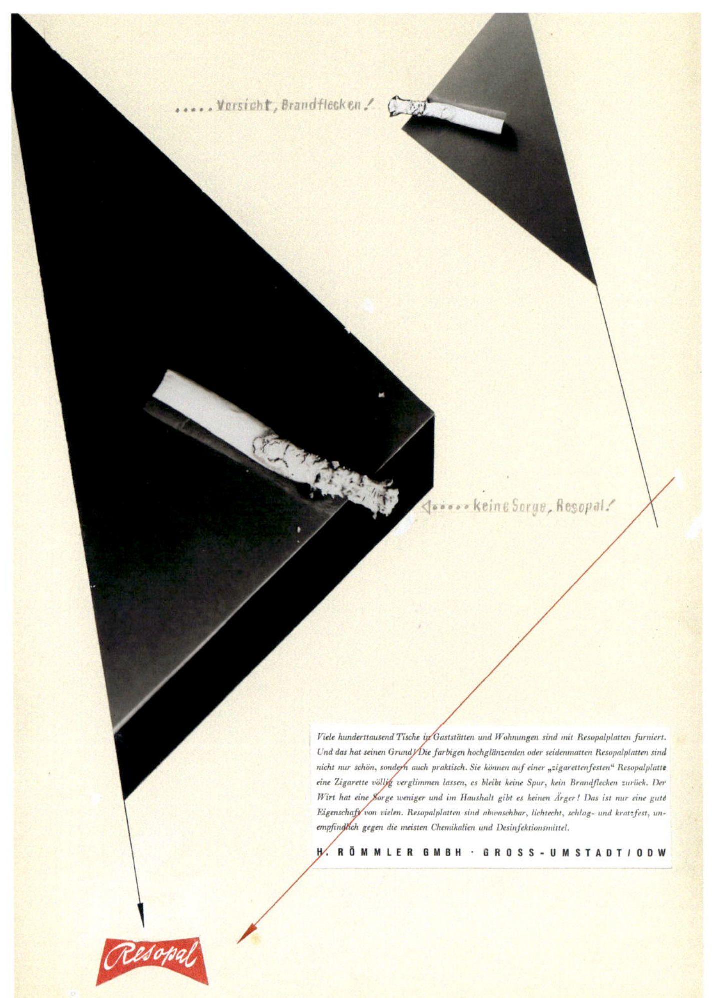

..... Versicht, Brandflecken!
..... keine Sorge, Resopal!
Viele hunderttausend Tische in Gaststätten und Wohnungen sind mit Resopalplatten furniert. Und das hat seinen Grund! Die farbigen hochglänzenden oder seidenmatten Resopalplatten sind nicht nur schön, sondern auch praktisch. Sie können auf einer „zigarettenfesten" Resopalplatte eine Zigarette völlig verglimmen lassen, es bleibt keine Spur, kein Brandflecken zurück. Der Wirt hat eine Sorge weniger und im Haushalt gibt es keinen Ärger! Das ist nur eine gute Eigenschaft von vielen. Resopalplatten sind abwaschbar, lichtecht, schlag- und kratzfest, unempfindlich gegen die meisten Chemikalien und Desinfektionsmittel.
H. RÖMMLER GMBH · GROSS-UMSTADT/ODW
Resopal

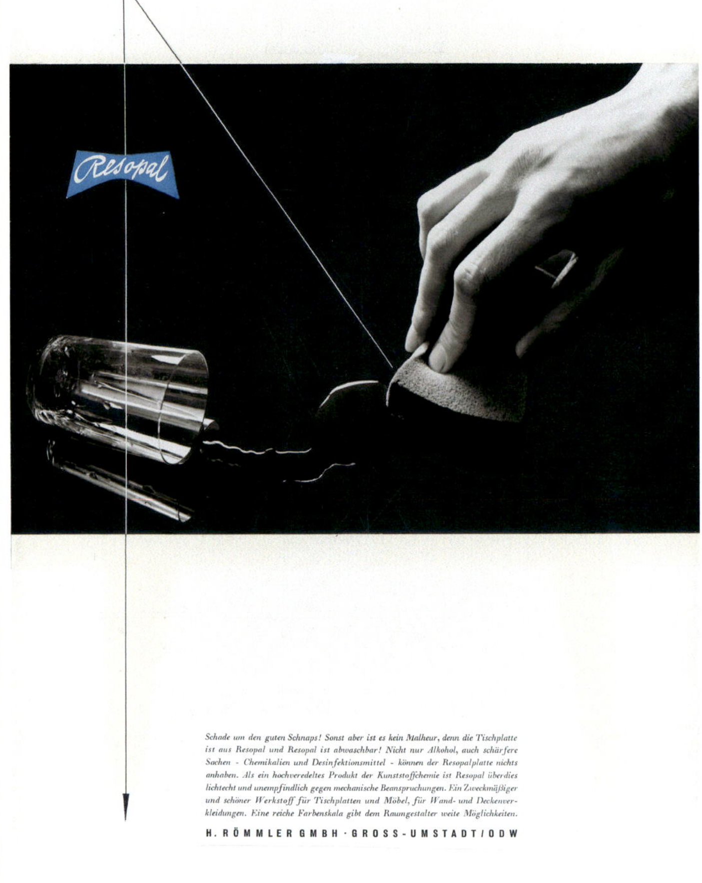

Resopal
Schade um den guten Schnaps! Sonst aber ist es kein Malheur, denn die Tischplatte ist aus Resopal und Resopal ist abwaschbar! Nicht nur Alkohol, auch schärfere Sachen - Chemikalien und Desinfektionsmittel - können der Resopalplatte nichts anhaben. Als ein hochveredeltes Produkt der Kunststoffchemie ist Resopal überdies lichtecht und unempfindlich gegen mechanische Beanspruchungen. Ein Zweckmäßiger und schöner Werkstoff für Tischplatten und Möbel, für Wand- und Deckenverkleidungen. Eine reiche Farbenskala gibt dem Raumgestalter weite Möglichkeiten.
H. RÖMMLER GMBH · GROSS-UMSTADT/ODW
Resopal für Tisch, Schrank und Wand

■ JUPP ERNST: WERBEMITTEL FÜR RESOPAL, VOR 1955

Wie schon bei seinen Anzeigen aus den dreißiger Jahren ist es wiederum der konstruktiv-architektonische Aspekt, der die Darstellung bestimmt und die unterschiedlichen Möglichkeiten weist, wie die Resopalplatte eingesetzt werden kann. Daneben werden die besonderen Eigenschaften des Materials wie helle, freundliche Farbigkeit, spiegelglatte Oberfläche, Unverwüstlichkeit und leichte Reinigung betont.

■ JUPP ERNST: ADVERTISING MATERIAL FOR RESOPAL, PRE-1955

As in his advertisements back in the thirties, the structural, architectural aspect once more dominates the graphics and demonstrates the range of different ways of using Resopal panels and boards. The material's particular features such as its light, cheerful colours, smooth surface, combined with the fact that it was indestructible and easy to clean, were also emphasised.

Farbkarte

Durch Unempfindlichkeit gegen Wasser, Fett und alle im Haushalt üblichen Lösungsmittel, Farben und Säfte ist Resopal der ideale Oberflächenbelag für Küchenmöbel. Abwischen mit Wasser und Seife genügt, um Tisch, Schrank und Regal wieder in der ganzen Schönheit der Resopal-Pastellfarben erstrahlen zu lassen.

grau 161

türkis 702

lindgrün 601

olivgrün 671

Leinen beige 3140

Leinen grau 3115

Leinen grün 3160

Leinen blau 3170

hellblau 720

rauchblau 750

mittelblau 742

dunkelblau 761*

weiß-gewolkt 1331

hellbraun 521

grün-gewolkt 1521

pompejanisch rot 1251

elfenbein 113

sandfarben 116

hellgelb 440

gelb 442*

Die mit * bezeichneten Farben sind nur im Alwine-Format ca. 800 x 1750 mm lieferbar, die Farbe rauchblau 750 nur im Claudia-Format ca. 1250 x 2800 mm, alle übrigen Farben im Alwine- und Claudia-Format.

Rohstoffbedingte Farbabweichungen vorbehalten!

Liane elfenbein 3210

Liane grau 3211

Liane olivgrün 3206

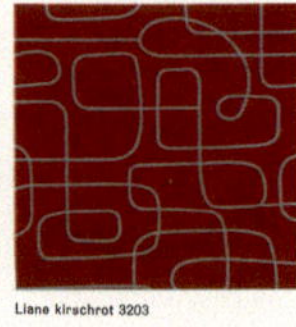

Liane kirschrot 3203

beige 120

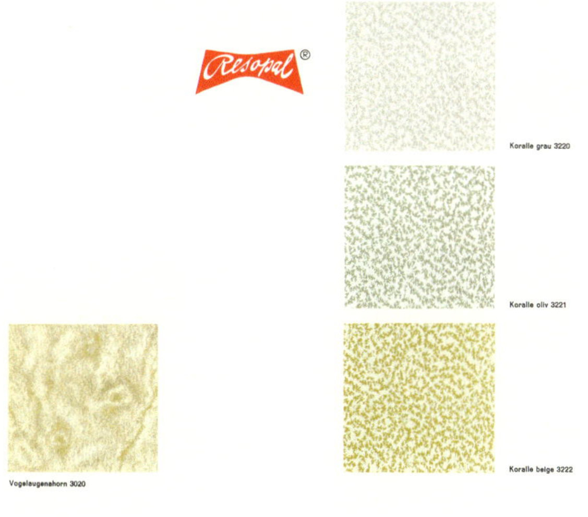

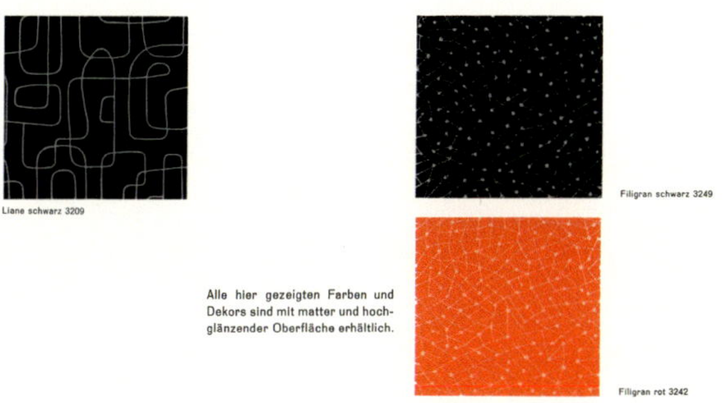

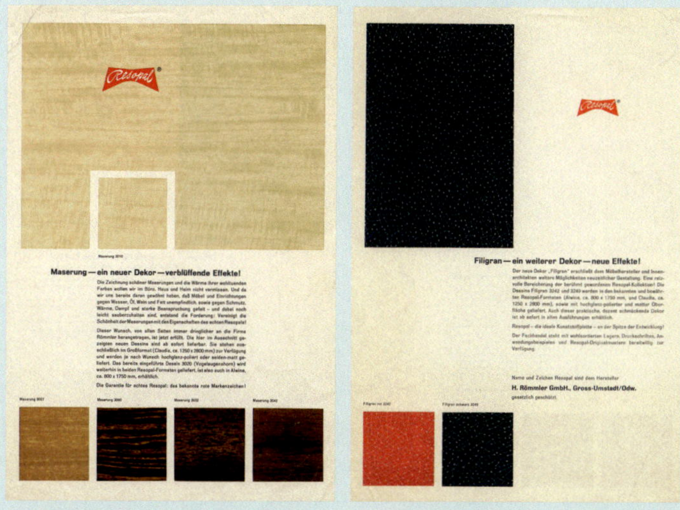

PROSPEKT »MASERUNG« UND »FILIGRAN«, 1959

Jupp Ernst trat dafür ein, nur solches Dekor zu entwickeln und einzusetzen, das »dem Kunststoff Eigenart gibt«. Jede Imitation von Holz, Marmor oder anderen Naturprodukten lehnte er vehement ab. Als er 1953 mit dem Entwurf eines Prospekts betraut wurde, in dem die Bezeichnungen »Birke geschält«, »Vogelaugenahorn«, »pompejanisch rot« und »mahagoni rot« vorkommen sollten, weigerte er sich und verwies auf das »feste Übereinkommen mit der Geschäftsleitung«, niemals in die Situation gebracht zu werden, für Surrogate werben zu müssen. Es scheint, dass er damit Erfolg hatte. Denn erst 1959 wirbt der Prospekt »Maserung« explizit für ein Holzdekor. Ironischerweise wird auf der anderen Seite des Werbeblatts das Dekor »Filigran« von Jupp Ernst angepriesen. Der doppelseitige Prospekt wurde von Wolfgang Schmittel gestaltet, der zu dieser Zeit bereits für die Braun AG tätig war und später die Corporate Identity dieser Firma für viele Jahre bestimmen sollte.

BROCHURE ON "WOOD GRAINS" AND "FILIGREE," 1959

Jupp Ernst believed that the only designs for Resopal that should be developed and used should be those that "give the synthetic material a quality of its own." He vehemently rejected any imitation of wood, marble or other natural products. In 1953, when he was asked to design a brochure that was to include the descriptions "birch veneer," "bird's-eye maple," "Pompeii red" and "mahogany red," he refused, referring to the "firm agreement with management," that he would never be put in the position of having to advertise surrogates. He seems to have been successful, because it was not until 1959 that a brochure on "wood grains" explicitly advertises a wood pattern. Ironically, the reverse side of the advertising leaflet features Jupp Ernst extolling the virtues of the design named "filigree." This double-sided flyer was designed by Wolfgang Schmittel, who was already working for Braun AG and would later dictate this company's corporate identity for many years.

◄ **WOLFGANG SCHMITTEL: PROSPEKT DER RESOPAL-KOLLEKTION, 1957**
◄ **WOLFGANG SCHMITTEL: BROCHURE FOR THE RESOPAL COLLECTION, 1957**

■ **AUSSTELLUNGSSTÄNDE MIT RESOPALPLATTEN, AB 1957**

Die H. Römmler GmbH stellte bei ihren Firmenauftritten die Farbigkeit der Resopalplatte in den Vordergrund. Für ihre Ausstellung bei der
Domotechnika setzte die Firma Braun ein Standsystem aus steinwei-
ßen Resopalplatten ein, das bis 1999 im Einsatz war.
1 Installation in Hamburg, 1958 **2** Messestand auf der Interzum, 1959
3 Frankfurter Frühjahrsmesse, 1957 **4** Domotechnika-Stand der Firma
Braun, 1986

■ **PRESENTATION FURNITURE AND ACCESSORIES MADE FROM
RESOPAL LAMINATES, FROM 1957**

In its stands for trade shows H. Römmler GmbH focused on the colour
possibilities offered by Resopal laminates. For its exhibit at Domotechnika, Braun developed a stand system in stone-white Resopal panels
that it continued to use until 1999.
1 Installation in Hamburg, 1958 **2** Stand for the Interzum show, 1959
3 Frankfurt Spring Fair, 1957 **4** Domotechnika stand for Braun, 1986

1

2

Resopal
für Küchen
für den Ladenbau
für Möbel
für sanitäre Anlagen
H. RÖMMLER GMBH, GROSS-UMSTADT/ODENWALD

3

Information International
BRAUN

4

Charakteristisch für die Milchbars der fünfziger Jahre war ihre Ausstattung mit Resopal. Aufnahme aus Düsseldorf, 1952
Resopal fittings were characteristic of the milk bars popular in the fifties. Photograph of a milk bar in Düsseldorf, 1952

RESOPAL IN ÖFFENTLICHEN RÄUMEN
RESOPAL IN PUBLIC SPACES

Nach ihrer Konsolidierung gelang der H. Römmler GmbH in wenigen Jahren die Erschließung eines großen Marktes für Resopalplatten, die schließlich viele Räume des modernen Lebens erobern sollten.

1956 waren bereits mehr als 120 verschiedene Dekore für Oberflächen im Angebot. Neben den bereits eingeführten Produkten wandte sich die Firma auch stetig neuen Fabrikationszweigen zu und stellte Großformteile in gepresster und gespritzter Ausführung, Möbel- und Bauelemente, Wandverkleidungen und mobile Trennwände her. Außerdem wurde die Produktion von Pressmassen wieder aufgenommen. Während bis 1951 für die Dekorschicht nur Harnstoffharz verwendet wurde, konnte mit dem Einsatz von Melaminharz, das eine größere Härte besitzt, die Qualität der Resopalplatte verbessert werden. Die Kollektion wurde durch neue Farben, neue reliefartige Strukturen für die Oberflächen sowie ein Material für die Außenanwendung ergänzt.

Insbesondere da, wo sich Resopal unauffällig, das heißt monochrom, modernen Objekten und Räumen anpasste, erschien es im besten Sinne modern. Kam Dekor ins Spiel, war Resopal als Synonym für Imitation heftig umstritten. Aber als das Unternehmen in den siebziger Jahren – inzwischen hieß es Resopal Werk H. Römmler GmbH – als Reaktion auf das wachsende Umweltbewusstsein anfing, Naturmaterialien nachzubilden, wollte man damit mehr als imitieren. Resopal sollte wie das echte Material wirken. Auf die Holzbilder wurden Porenstrukturen, auf Schiefer-Reproduktionen Bruchkanten, auf Stoffmuster Webstrukturen gepresst. Ein neuer Industriezweig, die Pressblechgravur entstand. Offensiv vertrat die Firma den Standpunkt: »Alle moralische Aggression, die von den Verarbeitern der traditionellen Materialien gegen die Kunststoffe geführt wird, ist Unsinn (...) wenn die neuen Materialien imstande sind, uns die gleiche Optik oder Anschauung zu liefern. Was spielt es schon für eine Rolle, welche Moleküle unsere Anschauung erzeugen?« Anfang der neunziger Jahre wurde Resopal in 26 verschiedenen Oberflächen, 214 Unifarben, 276 grafischen Dessins, 121 naturnahen Holz- und Gesteinsreproduktionen, ungezählten Farbdekorvarianten in der Overline-Technik und als Resopalor angeboten. Weil sich diese große Vielfalt jedoch als beliebig erwies, reduzierte die Resopal GmbH, wie die Firma nun hieß, die Angebotspalette etwa zehn Jahre später rigoros.

Within just a few years after its consolidation Römmler GmbH managed to open up a vast market for Resopal laminates, which would ultimately find their way into many spheres of modern life.

In 1956, over 120 different designs were already available for surface finishes. As well as its established products, the company constantly explored new areas of manufacturing and made large injection- and compression-moulded articles, furniture and building components, wall panelling and movable partitions. They also restarted production of moulding compounds. Whereas up to 1951 only urea resin was used to impregnate the decorative layer of paper, the use of the harder melamine resin made it possible to improve the quality of the Resopal laminate. New colours, new relief-like textures for the surface finishes and a product specially designed for outdoor use were added to the collection.

In particular in instances where Resopal was inconspicuous, i.e. monochrome, fitting in discreetly with modern buildings and spaces, it appeared modernist in the best sense. Whenever it had a decorative element, Resopal, as a synonym for imitation, was highly controversial. However, in the seventies when the company – which in the meantime had been renamed Resopal Werk H. Römmler GmbH – began to make reproductions of natural materials as a response to the public's growing environmental awareness, the ambition was no longer merely to imitate: Resopal was intended to actually appear to be the real thing. Pore structures were moulded onto the wood images, broken edges onto reproduction slate, woven structures onto fabric patterns. A new industry for special press plates was born. To counter its opponents, the company went on the offensive, saying: "All moral aggression directed against synthetics by people who work with traditional materials is nonsense (...) if the new materials are able to give us the same optical quality or image. What difference does it make which molecules produce the image we perceive?" In the early nineties, Resopal was available in 26 different surface finishes, 214 plain colours, 276 graphic designs, 121 authentic looking reproduction wood and stone effects, countless design options with the help of the so-called overline-technique and as Resopalor. However, about ten years later this huge variety turned out to be arbitrary, so that Resopal GmbH, as the company is now called, reduced its range rigorously.

4

Sie wurden bereits um 1938 eingesetzt und waren in der Nachkriegszeit in beiden deutschen Staaten – in der DDR unter dem Namen
Sprelacart – weit verbreitet.

1 Straßenbahn mit Resopaldecke, sechziger Jahre **2** Vorortzug mit Resopal-Wandverkleidung, sechziger Jahre **3** Berliner S-Bahn-Wagen mit
Sprelacart-Wandverkleidung, bis 2000 **4** Neue Resopalausstattung für
die erste ICE-Generation, 2006

Resopal panelling was used as early as 1938 and was widespread in
the post-War era in both German states. In East Germany it was called
Sprelacart.

1 Tram with Resopal ceiling, 1960s **2** Suburban train with Resopal wall
panelling, 1960s **3** Berlin S-Bahn carriage with Sprelacart wall panelling, up to 2000 **4** New Resopal interiors for the first generation of ICE
trains, 2006

■ **RESOPAL FÜR VERKEHRSBAUTEN**

1 Beim Elbtunnel in Hamburg ist der Versorgungskanal mit Resopalplatten verkleidet, die sich bei Bedarf leicht abnehmen und wieder
anbringen lassen. **2** Haltestelle in Hamburg
Beide Fotos stammen von 1975.

■ **RESOPAL FOR TRANSPORT STRUCTURES**

1 The services duct in the Elbe tunnel in Hamburg is clad with Resopal
panels that can be easily removed and re-fitted. **2** Tram stop in Hamburg
Both photos are from 1975.

1

2

Verdener Straße
H

■ **RESOPAL IN LÄDEN**

Für deren Innenausstattung waren lange Zeit monochrome Flächen oder Holzdekore bestimmend. Auch Pressteile aus Resopal wie der abgebildete Zahlteller gehörten dazu. Seit einigen Jahren sind unter den vielen Auswahlmöglichkeiten auch Metalloberflächen in Resopal möglich. Ein Beispiel dafür ist das WMF Kommunikationszentrum in Geislingen/Steige, 2003.

■ **RESOPAL IN SHOPS**

For a long time, monochrome surfaces or wood dominated shop fittings. Moulded elements in Resopal, such as the counter mat depicted here also featured. For a number of years, the range of available options has also included Resopal-laminated metal surfaces in Resopal. An example of this is the WMF communications centre in Geislingen/Steige, 2003.

1

2

3

4

■ RESOPAL IN GASTSTÄTTEN

1 Jupp Ernst: Anzeige, vor 1955 **2** Anzeige aus den sechziger Jahren
3 Tablett aus Resopal-Pressmasse **4** Kantinengeschirr aus Resopal-
Pressmasse **5** Wandbild im Speisesaal einer Kurklinik, 1974 **6** Auto-
bahnraststätte, siebziger Jahre

■ RESOPAL IN CATERING FACILITIES

1 Jupp Ernst: Advertisement, pre-1955 **2** Advertisement from the 1960s
3 Tray in moulded Resopal **4** Canteen tableware in moulded Resopal
5 Mural in the dining hall of a health spa clinic, 1974 **6** Motorway service station, 1970s

5

6

1 Hans Oberhoff: Wandbild im Hallenbad der Stadt Montabaur, 1967. 1996 wurde es im Zuge einer Modernisierung entfernt. **2** Hans Oberhoff: Dekor »Pyramide« aus der Kollektion »Serienunterdrucke« in den Farben des Wandbildes im Hallenbad Montabaur **3** Hallenbad Wörth, siebziger Jahre **4** Hallenbad Bad Neuenahr, siebziger Jahre **5** Kiteboard der Firma APK GmbH, »Digital in Resopal«, 2005

1 Hans Oberhoff: Mural in the municipal swimming pool, Montabaur, 1967. It was removed in 1996 as part of a modernisation scheme.
2 Hans Oberhoff: "Pyramid" pattern from the " Serienunterdrucke" collection in the colours of the mural at Montabaur swimming pool.
3 Wörth swimming pool, 1970s **4** Bad Neuenahr swimming pool, 1970s
5 Kiteboard of APK GmbH, "Digital in Resopal", 2005

3

4

5

1

■ RESOPALOBERFLÄCHEN BEI BÜROMÖBELN

1 Helmut Lortz: Gestaltung eines Tisches für eine Druckerei, sechziger Jahre **2** Der Belag des Schreibtisches aus den vierziger Jahren imitiert die damals gängige, grüne Linoleumauflage. **3** Das Möbel wurde 1964 von Rudolf Lübben entworfen, der mit Vorliebe weißes Resopal für Beläge und Fronten verwendete.

■ RESOPAL SURFACES FOR OFFICE FURNITURE

1 Helmut Lortz: Design for a work table for a printing works, 1960s **2** The surface of the desk from the forties imitates the green linoleum common at the time. **3** The furniture was designed in 1964 by Rudolf Lübben, who liked to use white Resopal for the tops and fronts of his furniture.

2

3

1 Schrankwand im Büro **2** Großraumbüro der Firma DEMAG, Wetter/
Ruhr **3** Holzdekor in einem Vorzimmer **4** EDV-Zentrale bei der Bundesanstalt für Arbeit, Nürnberg **5** Schaltzentrale bei der Bundesanstalt
für Arbeit, Nürnberg
Alle Aufnahmen stammen aus den siebziger Jahren.

1 Wall unit in an office **2** Open-plan office at DEMAG, Wetter/Ruhr
3 Wood pattern Resopal in a secretary's office **4** IT centre at the Federal Labour Office (BfA), Nürnberg **5** Control room at the Federal Labour
Office(BfA), Nürnberg
All the photographs are from the 1970s.

5

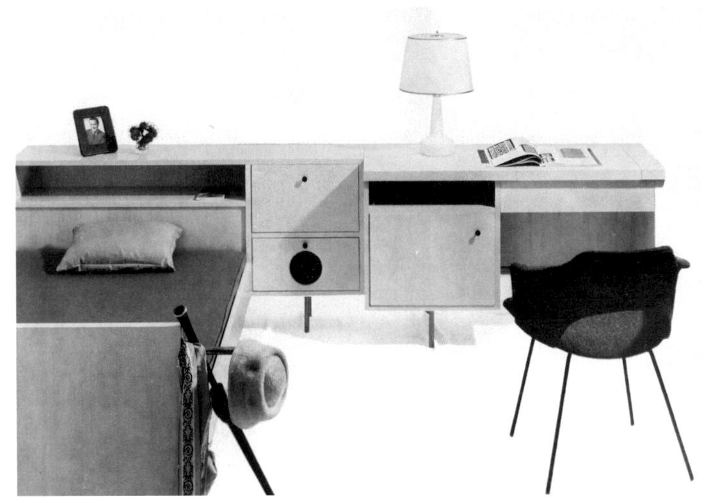

■ RUDOLF LÜBBEN: HOTELMOBILIAR AUS RESOPAL

1 Zimmereinrichtung, Hamburg, 1952 **2** Hotelgerätewagen, 1956 **3** Hotel »Tropicana«, Lomé (Togo), 1970. Die Einrichtung in Resopal erfolgte wegen der exzellenten Tropentauglichkeit. **4** Sesselgruppe mit Resopalablagen, 1960 **5** Polstermöbelsystem mit Resopalablagen, 1960

■ RUDOLF LÜBBEN: HOTEL FURNISHINGS IN RESOPAL

1 Room furnishings, Hamburg, 1952 **2** Hotel cleaner's trolley, 1956 **3** Hotel "Tropicana", Lomé (Togo), 1970. Resopal was chosen for the furniture and fittings because it is highly suitable for tropical conditions. **4** Seating with integrated drinks rests in Resopal, 1960 **5** Modular reception seating with integrated Resopal coffee tables, 1960

4

5

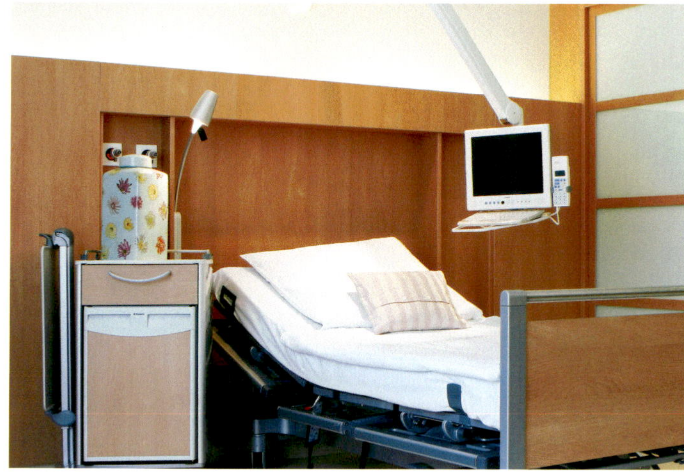

■ RESOPALMÖBEL IN KULTUR- UND BILDUNGSEINRICHTUNGEN

1 Tresen Bücherei Ütersen, fünfziger Jahre **2** Wandregale in einem Kindergarten, siebziger Jahre **3** Foyer des Hessischen Landesmuseums in Darmstadt um 1966. Der Empfangstresen wurde mit Resopal perlweiß belegt. **S. 85** Spanische Schulbank, um 1960

■ RESOPAL FURNITURE IN CULTURAL AND EDUCATIONAL FACILITIES

1 Library counter, Ütersen, 1950s **2** Wall shelving in a kindergarten, 1970s **3** Foyer of the Hessisches Landesmuseum in Darmstadt around 1966. The reception counter was finished in pearl-white Resopal. **S. 85** Double school desk and integrated chairs, around 1960

■ KÜNSTLERISCHE GESTALTUNG MIT RESOPAL

1 HAP Grieshaber: Wandbild in der Mensa der Johann Wolfgang Goethe-Universität, Frankfurt/Main, um 1960 **2** Bernd Krimmel: Gestaltung der Klassenzimmertüren im Geschwister-Scholl-Schulzentrum von Novotny Mähner Architekten, Groß Krotzenburg, 1966 **3** Armin Wermann: Wandgestaltungen im Schulzentrum Kusel, 1982

■ RESOPAL IN ARTISTIC DESIGN WORK

1 HAP Grieshaber: Mural in the student refectory at Johann Wolfgang Goethe University, Frankfurt/Main, around 1960 **2** Bernd Krimmel: Pictures on classroom doors at the Geschwister Scholl School Centre, by Novotny Mähner Architekten, Gross Krotzenburg, 1966 **3** Armin Wermann: Murals at the Kusel School Centre, 1982

■ **CAROLA SCHÄFERS ARCHITEKTEN: SEMINARZENTRUM,**
FREIE UNIVERSITÄT BERLIN, 2006

Bei dem Umbau einer früheren Cafeteria sind Büros und drei Seminar-
räume entstanden, von denen der innen liegende Seminarraum durch
seine Farbe und Materialität besonders hervorgehoben wurde. Charakteristisch ist seine zum Flur hin großflächige, nahezu fugenlose
Wandverkleidung aus Resopalpaneelen. Das Streifendesign wurde vom
Architekturbüro selbst entworfen und im Digitaldruck umgesetzt. Das
Dekor »Paprika« in der Teeküche stammt aus der aktuellen Kollektion.

■ **CAROLA SCHÄFERS ARCHITEKTEN: SEMINAR CENTRE,**
FREIE UNIVERSITÄT BERLIN, 2006

A cafeteria was converted creating offices and three seminar rooms.
The seminar room in the interior of the building is given special
emphasis through the use of colour and materials. The large expanse
of Resopal panels with almost no joints on the wall to the corridor is a
dominant feature. The architects designed the striped pattern themselves, which was converted into a digital print. The "Paprika" design in
the tea kitchen is from the current collection.

■ **CAROLA SCHÄFERS ARCHITEKTEN: CAFETERIA GARYSTRASSE, FREIE UNIVERSITÄT BERLIN, 2006**

Ziel war es, für die im Souterrain liegende Cafeteria neben einer besseren Belichtung und einem neuen Farbkonzept eine Einrichtung zu entwerfen, die auch außerhalb der Öffnungszeiten den Studenten als Arbeitsraum und für Veranstaltungen zur Verfügung steht. Sowohl die Schiebetüren, die den Gastraum von der Essensausgabe trennen, als auch die Lüftungskanäle wurden mit Resopal in der Farbe Aubergine verkleidet. Für die neu angefertigten Tische und Bänke wurde das Dekor »Noah Teak« verwendet.

■ **CAROLA SCHÄFERS ARCHITEKTEN: CAFETERIA ON GARYSTRASSE, FREIE UNIVERSITÄT BERLIN, 2006**

The aim was to create better lighting, a new colour scheme and furniture and fittings for a basement cafeteria that students can use outside opening hours as a room to work in or hold events. Both the sliding doors that separate the table area from the food counter and the ventilation ducts are finished in aubergine coloured Resopal panels. The "Noah Teak" design was used for the new tables and benches.

Eva Brachert

LEBEN UND WOHNEN MIT RESOPAL NACH DEM ZWEITEN WELTKRIEG
LIVING WITH RESOPAL AFTER THE SECOND WORLD WAR

»Wie unser deutsches Wort ›Hausfrau‹ sagt, besteht ein unmittelbarer (…) Zusammenhang zwischen Frau und Haus. Trotz der neuzeitlichen, durch Raumknappheit und -verteuerung spürbar erschwerten Bedingungen muss man sich verantwortungsbewusst um die Entwicklung einer zeitgemäßen Wohnkultur bemühen«[1], so lautete der Anspruch an die Bundesbürgerin in den fünfziger Jahren.

Der nach Ende des Krieges beschrittene demokratische Weg hatte zweifellos auch Folgen für Lebensgefühl und Wohnkultur, die sich in der Gestaltung von Wohnraum, Möbeln und Alltagsgegenständen offenbarten. Nie zuvor hatte die Wohnungseinrichtung eine wichtigere gesellschaftliche Rolle gespielt, wurde die Gestaltung des persönlichen Lebensraums mit individuellen Lösungen in solchem Maße öffentlich thematisiert.

Doch welche Formen entsprachen der neuen Gesellschaft? Die »Ratgeber in Geschmacksfragen« hatten in der gesellschaftlichen Umbruchsphase der frühen Nachkriegsjahre zunächst Schwierigkeiten, sich gegen die noch immer vorherrschende traditionelle Ordnung des Lebensstils durchzusetzen. Denn entgegen der erfahrenen Zerstörung lebte der Traum vom gutbürgerlichen Heim weiter, das sich in festgeschriebenen Raumfolgen wie Raucherzimmer, Damensalon, Musikzimmer und in ganz bestimmten Ausstattungsgegenständen wie Büfett, Spieltisch und Flügel darstellte. Trotz verschiedener Reformbemühungen, durch formale Klarheit und reine Funktionalität in der Einrichtung einen neuen Wohnstil durchzusetzen, ließ die Möblierung noch vorrangig den Gedanken an das »gemütliche Heim« mit unfunktionalen, aber repräsentativen Einrichtungsgegenständen erkennen. Schwere Möbel und großgemusterte Dekorstoffe sollten die wohlgeordnete bürgerliche Gesellschaftshierarchie fortleben lassen. Kunststoffgegenstände hatten hier zunächst noch keinen Platz. Sie gehörten einfach nicht in die repräsentativ ausgerichteten Wohnwelten, auch wenn sie die Organisation des Lebens erleichtern konnten.

Dass sich die alten gesellschaftlichen Strukturen zwar in Auflösung befanden, aber im privaten Rahmen noch gelebt wurden, zeigte sich auch in der immerhin propagierten, aber faktisch nicht umgesetzten Gleichberechtigung der Geschlechter. Mit welchem Frauenbild wurden unsere Mütter und Großmütter damals durch Wirtschaft und Medien konfrontiert? Krieg und Nachkriegszeit waren überstanden, eine Zeit, in der viele Frauen allein auf sich gestellt das Leben und den Alltag der Familie organisieren mussten. Und doch beherrschte weiterhin das traditionelle Frauenbild die öffentliche Meinung. Zeigte Werbung, vor allem die für Wohnungseinrichtungen, partnerschaftliche Geschlechterverhältnisse, wurden die alten Rollen reproduziert und gleichzeitig das hohe Lied auf die Küche als weiblicher Arbeitsplatz gesungen.[2] Ab etwa Mitte der fünfziger Jahre – der wirtschaftliche Aufschwung erfasste mehr und mehr Bereiche der Republik – schien jedoch ein neues Frauenbild zu entstehen: die innovationsfreudige, eigenständige, damenhafte Kundin. Dieser neue Typus wurde der bundesdeutschen Hausfrau in Werbung und Berichten vorgestellt und dabei häufig mit der Nutzung von Kunststoffen in Verbindung gebracht.

»Hat Monika nicht recht?« war der Titel einer Fortsetzungskolumne in der Zeitschrift *Kunststoff-Berater*. In diesem monatlich erscheinenden Magazin bemühten sich chemische Industrie, Handel und Handwerk seit 1950, umfangreiche Informationen über diesen Werkstoff an möglichst viele Konsumenten, Händler und Verarbeiter weiterzugeben. Und wo konnte diese Wissensvermittlung am nachhaltigsten beginnen? Im »Reich der Frau«[3], das entsprechend der neuen Wohnkultur nicht mehr Arbeitsküche sein sollte, sondern eine behagliche, individuelle Wohnküche.

Cornelius, der Erzähler, begleitet den Leser gemeinsam mit Monika, seiner neuen Freundin, durch das Jahr 1957. Dabei wirbt er für die modernen

"As our German word 'housewife' says, there is a direct (…) connection between women and the home. Despite the difficult conditions we are now facing as a result of living space being expensive and in short supply, we have to take a responsible attitude to developing a style of home decor that is appropriate for our time."[1] This was what was asked of German women in the 1950s.

The democratic path Germany followed after the War without doubt had an effect on the way people felt about life in general and about their home life, which was reflected in the design of living spaces, furniture and everyday objects. Never before had home decor played such an important role in society, nor had the individualized design of personal living space been so much in the public eye.

However, what forms were appropriate for this new society? In the transition in society that occurred during the early post-War years, the "advisors on matters of taste" had some difficulty in asserting themselves against the traditional ideas on lifestyle that still prevailed. At odds with the reality of the destruction people were actually living with, the dream of the good bourgeois home lived on and manifested itself in prescribed sequences of rooms, such as the smoking room, ladies drawing room, music room, and in very specific articles of furniture such as the sideboard, card table or grand piano. Despite various reform endeavours to introduce clarity of form and pure functionality into furnishings and establish a new style of home decor, the furnishing style that prevailed showed that people were thinking primarily in terms of the "cosy home" with its non-functional but showy furnishings. Heavy furniture and fabrics with loud patterns were intended to ensure the survival of the well-ordered bourgeois social hierarchy. Plastic articles had no place here. They simply did not belong in prestigiously furnished homes, despite the fact that they could have made it easier for people to organise their lives.

Although old social structures were in a process of dissolution, they were still alive and kicking in the private realm. This could also be seen in the question of equality of the sexes, which was being publicly proclaimed but not actually put into practice. What image of women were industry and the media trying to sell to our mothers and grandmothers? They had survived the War and the post-War years, a time in which many women had had to organise their lives and cope with the practicalities of family life completely alone. Yet the traditional image of women still dominated public opinion. Whenever advertising, particularly for home furnishings, depicted couples, the old roles were wheeled out and the kitchen eulogized as the woman's workplace.[2] However, from about the mid-1950s, as the economic boom affected more and more areas of life in West Germany – a new stereotype of woman started to emerge: the innovative, independent and ladylike customer. This new type of woman was introduced to West German housewives in advertising and media reports and frequently associated with the use of plastics.

"Is Monika not right?" was the title of a regular column in the magazine *Kunststoff-Berater*. In this monthly magazine, the title of which translates as "Plastics Advisor," the chemicals industry, along with trade and commercial associations, had been working hard since 1950 to disseminate comprehensive information about this material to as many consumers, retailers and processing companies as possible. What was the best place for this transmission of knowledge to begin? In the "woman's realm,"[3] the kitchen, which in line with the new trends in home decor, was no longer meant to be merely a place to work in, but also a comfortable and individualized living space.

Accompanied by his new girlfriend Monika, Cornelius, the narrator, takes the reader through the year 1957. All the while, he is disseminating a mes-

H. Römmler GmbH: Werbung für die Reso-
palküche, fünfziger Jahre
Advertisement for the Resopal kitchen by
H. Römmler GmbH, 1950s

Werbung der Firma Tielsa, fünfziger Jahre
Advertisement by Tielsa, 1950s

gesellschaftlichen Entwicklungen: Ansporn für innovative Leistungen des Mannes seien die Forderungen der Frauen nach verbesserten Arbeitsbedingungen. Die wirtschaftlich und praktisch denkende Monika lotst den sich lange widersetzenden Cornelius mit List in ein Kücheneinrichtungsgeschäft, wo er durch die »Helligkeit und Farben, [einer in] Glanz und Gloria leuchtenden Küche« enorm beeindruckt ist. Vor dem Hintergrund dieses »wahren Triumphs der Küchentechnik« erklärt sie ihm, dass nur mit einer solchen Küche ihre Ehe einen Anfang nehmen könne. »Und während ich noch ungläubig starrte, wurde ich mit einem wahren Feuerwerk von neuen, unbekannten Begriffen überschüttet, von einer geschäftig und glücklich demonstrierenden Monika hin- und hergezerrt. Ich lernte, was eine Kunststoffplatte ist und welche unzweifelhaften Vorzüge sie besitzt.«[4] Ziel dieser inszenierten Wirklichkeit war, Leserin und Leser in einen soziokulturellen Kontext einzuführen, der mit Kunststoffen verbunden wurde. Vor allem galt es, aus Käuferinnen Konsumentinnen zu machen, also bei ihnen die Bereitschaft zu wecken, auch etwas über den unmittelbaren Bedarf hinaus zu kaufen. Die Kunststoffwerbung, die ihre Produkte geschickt als Metapher einer erstrebenswerten neuen Wohnkultur darstellte, suggerierte eindrücklich, dass nur pflegeleichte, glatte Kunststoffoberflächen der Frau den unvorstellbaren Spagat zwischen der erträumten uneingeschränkten persönlichen Flexibilität mit einem Höchstmaß an Komfort *und* dem Einsatz zum Wohle der Familie ermöglichen. Freie Zeit trotz Haushalt und Berufstätigkeit, das waren neue gesellschaftsspezifische Ausprägungen, die zu Änderungen nicht nur im Rollenverhalten führten, sondern auch nach Lösungen in der Wohnkultur verlangten.

Und tatsächlich entwickelte sich mit dem Konjunkturaufschwung im Land ein neues Konsumverhalten. Die in Massenanfertigung preiswert produzierten Kunststoffartikel ließen sich zum einen in die noch traditionelle Gesellschaftsstruktur integrieren, zum anderen verbanden sie sich allmählich im kollektiven Bewusstsein mit dem neuen demokratischen Lebensstil. Für die Küche hieß das, dass dieser Raum durch erweiterbare Anbausysteme, je nach persönlichen finanziellen Mitteln, zu einem wohnlichen Ambiente ausgebaut werden konnte.[5] Auf diese Weise kam über den in Funktion und Stellenwert neu definierten »Arbeitsplatz der

sage about the modern developments taking place in society: the incentive for men's innovative achievements – was the underlying suggestion – was women's demands for better working conditions. Thrifty and practically minded Monika craftily steers Cornelius after much resistance into a kitchen showroom, where he is hugely impressed by the "brightness and colour [of] a kitchen in all its splendour". Against the backdrop of this "true triumph of kitchen technology," she explains to him that she can only contemplate embarking on married life if she has a kitchen like this. "And, as I still stood staring in disbelief, I was overwhelmed by an absolute barrage of new and unfamiliar terms, and then dragged all over the shop by Monika who was bustling around in her element. I discovered what a laminated worktop is and what indisputable advantages it has."[4]

The aim of this carefully orchestrated little scene was to introduce readers to a socio-cultural context that was closely related to plastics. The prime aim was to transform women from mere purchasers of items to satisfy immediate needs into consumers, i.e. to awaken their willingness to buy something beyond the purely utilitarian. Advertising that cleverly depicted plastic products as the metaphor for a desirable new lifestyle, subliminally suggested that only smooth easy-care plastic surfaces would make it possible for a woman to achieve the inconceivable juggling act of combining the personal unrestricted flexibility she dreamt of with the highest degree of comfort and a commitment to the welfare of the family. Leisure time – combined with going to work and running the home – these were features of a new kind of society, which not only led to changes in gender-role behaviour but also called for different solutions in home decor.

New consumer behaviour really did develop hand in hand with the economic boom in the country. Firstly, plastic cheaply mass-produced articles were integrated into the still traditional structure of society and secondly they gradually became associated in the collective consciousness with the new democratic way of life. This meant that add-on modular systems to suit the individual's pocket could be used to extend the kitchen and create a comfortable homely atmosphere.[5] In this way, a symbol of the new way of life gradually entered every West German

Werbung der Firma Gruco, fünfziger Jahre
Advertisement by Gruco, 1950s

Resopalwerbung »Modell: Herrenzimmer«,
um 1960
Resopal advertisement "Model: study,"
around 1960

Frau« mit den Kunststoffflächen in zarten Farbkombinationen ein Symbol des neuen Lebensstils nach und nach in jeden bundesdeutschen Haushalt. Dabei wurde die noch auf rationelle Haushaltsführung ausgerichtete Küche der zwanziger Jahre von einem Arbeitsraum zu einem zusätzlichen, freundliche Wohnlichkeit ausstrahlenden Lebensraum – »dem Wunsch jeder Hausfrau«[6]. Nun konnte sich auch hier das Familienleben abspielen, denn die Hausarbeit war wegen der pflegeleichten Kunststoffgegenstände ja schnell erledigt.

Beim Angebot dekorativer Schichtstoffplatten für Arbeitsflächen, Tische und Schrankfronten spielten die verschiedenen Marken zunächst noch keine große Rolle. Neben dem Material Resopal verwendeten die verschiedenen Möbelhersteller auch Produkte aus Pollopas, Ultrapas und Formica, einem amerikanischen Fabrikat.

Die dekorative Schichtstoffplatte, mit der die neuen Möbel belegt waren, unterstützte die Weiterentwicklung tradierter Gestaltungs- und Gesellschaftsformen, zumindest im Wohnbereich. Die in der Gliederung des Wohnraums vorstrukturierte traditionelle Welt mit ihren manifesten Rollenbildern konnte durch wandelbare Anbau- oder Systemmöbel zumindest variiert werden. Das Schlafzimmer wurde, indem man das Bett hochklappte und in der Schrankwand verschwinden ließ, zum Wohnzimmer mit Beistelltisch und Kunststoffstühlen. Alles war leicht zu heben und ließ sich für jede Lebenssituation neu gruppieren. So etwas erlaubten nur Möbel, die in ihrer Gestaltung nicht einem bestimmten Wohnraum zugeordnet waren, sondern als alleinigen Schmuck ihre seidenmatt glänzende, farbige Oberfläche boten, die in jedem Raum zur Steigerung der Wohnlichkeit beitragen sollten.

Auf diese Weise entwickelte sich in den ersten fünfzehn Nachkriegsjahren ein neuer Lebens- und Wohnstil, der sich dadurch auszeichnete, dass anstelle von Statussymbolen für wenige nun praktikable, individuelle Lebenslösungen für alle umsetzbar schienen. Diese Zeit trug auch den Namen »Kunststoffzeitalter«, denn nun waren es die Kunststoffwaren, die den materiellen und ideellen Alltag bestimmten. Einen wesentlichen Beitrag zur neuen Lebensart leistete der Werkstoff Resopal – und zwar als Schichtstoffplatte für die Möbelindustrie und als Pressmasse für die Gehäuse von modernen Elektrogeräten.

home via the "woman's workplace" that with its laminated surfaces in delicate colour combinations had been redefined both in terms of function and status. In the process, the kitchen of the 1920s, designed to facilitate rational housekeeping, was transformed from a workspace to an additional living space that radiated friendly homeliness – "every housewife's wish."[6] Now family life could centre on this space because the housework could be quickly got out of the way thanks to the easy-care plastic articles.

Initially the different brands were not relevant in the range of decorative laminates that were used for work surfaces, tables and cabinets. Apart from Resopal, different furniture manufacturers also used products made of Pollopas, Ultrapas and Formica, an American brand.

The decorative laminates that were used to finish modern furniture promoted the evolution of traditional forms of design and social mechanisms, at least in the home sphere. The traditional pre-structured world with its stereotypical gender roles, which was reflected in the organisation of living space, could now at least be varied through flexible add-on or modular furniture. The bedroom with a bed that folded up out of sight into a built-in wall unit could be turned into a living room with an occasional table and plastic chairs. Everything was easy to lift and could be reorganised to suit different living situations. That was only possible with furniture that did not obviously belong to a particular room by virtue of its very design, but with its satin-finish coloured surfaces provided the only ornament and was designed to enhance the homeliness of any room.

Thus in the first fifteen years after the War, a new way of life and style of living developed, characterised by the fact that instead of status symbols for the few, practicable, individual solutions to suit everyone's life seemed feasible. This period was known as the "plastics age" because plastic goods dominated everyday life – both practically and ideologically. Resopal made an important contribution to this new way of life – mainly through laminated board used in the furniture industry or moulding compound used to make casings for modern electrical appliances.

After 1950, industry tried to stimulate the public's growing interest in plastic products through advertising that suggested that their quality,

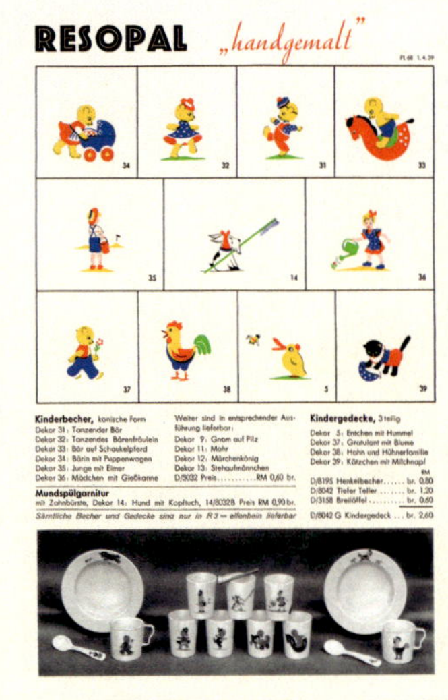

Prospekt für Resopalartikel, Plastica, 1934
Brochure for Resopal articles, Plastica, 1934

Marmeladendose mit Kunststoffteilen, dreißiger oder fünfziger Jahre
Jam dish, in glass and plastic, 1930s or 1950s

Die Industrie forcierte nach 1950 das wachsende Interesse an Kunststoffprodukten, indem sie propagierte, dass ihre Güte, die Qualität ihrer Verarbeitung und ihre moderne, ansprechende Form einen positiven Lebensstil bewirke. Bei vielen modern eingestellten Konsumenten überwog jedoch zunächst die negative Konnotation der ersten Kunststoffwaren aus den dreißiger Jahren, die in ihrer Gestaltung zumeist eine reine Imitation von handwerklich erzeugten Haushaltswaren darstellten.[7] Passend zur Ausstattung der deutschen Haushalte der späten dreißiger und vierziger Jahre mit dunklen Dekostoffen, massivem Hausrat, schweren Sesseln und Möbeln aus dunklem Holz, war auch das Design von Hausrat – selbst wenn aus Kunststoff hergestellt – auf das Niveau volkstümelnden Kunstgewerbes herabgesunken. Für Tassen, Salatbestecke, Eierbecher, Salzfässchen und Untersetzer aus Kunststoff zählte, selbst bei Ablehnung jeglicher Schmuckform, nur die Unverwüstlichkeit der Aminoplaste – der seinerzeit modernsten Kunststoffart, zu der auch Resopal zählte. Messeberichte wurden zum Beispiel mit Abbildungen von Zucker- und Marmeladendosen illustriert, die kleinmütig nur in vertrauter Formensprache umgesetzt waren, ohne Gefühl für die wahren Qualitäten des neuartigen Materials.[8]

Die ersten Pressstoffwaren nach dem Krieg waren ebenfalls spröde und steife, jedoch preiswerte Kleinartikel.[9] Damit standen der Hausfrau in den fünfziger Jahren im Unterschied zu den dreißiger Jahren zunächst keine wesentlich neu gestalteten Kunststoffwaren zur Verfügung. Das erklärt zum Teil deren zunächst zögerliche Durchsetzung im Alltag. Auch die Ausstellung »Wie wohnen« 1949 im Stuttgarter Landesgewerbeamt, auf der erstmals nach dem Krieg von zwei Firmen Hausrat aus Plastik präsentiert wurde, stand noch ganz im Zeichen nicht nur einer wirtschaftlichen, sondern auch einer designgeschichtlichen Kontinuität von Vor- und Nachkriegszeit. Bemerkenswert war jedoch die Präsentation der dekorativen Schichtstoffplatte der Venditor Kunststoff Gesellschaft, der auch die Dynamit AG angehörte – seit 1930 schärfste Konkurrentin der H. Römmler AG in der Herstellung und Verarbeitung von Pressstoffen. Auf diese erste Schau einer neuen Lebensart unter neuen Lebensbedingungen folgten Wanderausstellungen, die den Einfluss der Raumgestaltung auf eine leichtere und rationellere Bewältigung der Hausarbeit erläuterten und

high standard of manufacture and modern attractive design would create a desirable lifestyle. However, for many modern-minded consumers the negative connotations of the early plastic goods of the 1930s, which in design terms were usually a mere imitation of handcrafted household goods, still prevailed.[7] In keeping with the decor in German homes in the late thirties and forties with their dark fabrics, solid household goods, heavy armchairs and furniture in dark wood, the design of household goods – even when they were made of plastic – had sunk to the level of folksy arts and crafts. For cups, salad servers, egg cups, salt cellars and saucers in plastic, what counted, even if all form of ornament was rejected, was solely the indestructible nature of the aminoplasts – at the time the most modern type of plastic, which also included Resopal. Even trade show reports, for example, were illustrated with pictures of sugar bowls and jam dishes that had been designed slavishly using the familiar formal language without any feeling for the true qualities of this new material.[8]

The first moulded articles after the War were also rather plain and uninspired, albeit cheap.[9] So unlike in the 1930s, the plastic articles the German housewife of the fifties had to choose from did not have any essentially new design features. That explains to some extent why they took so long to become an established part of everyday life. Even the exhibition "Wie wohnen" held in 1949 in Stuttgart's *Landesgewerbeamt* or trading standards office, at which for the first time since the end of the War two companies exhibited household goods made of plastic, was still characterised by continuity between the pre-War and post-War years – not only in economic but also in design terms. However, what was significant was the company Venditor Kunststoff Gesellschaft's exhibit of decorative laminates. This company also belonged to Dynamit AG, which since 1930 had been H. Römmler AG's toughest competitor in the production and processing of moulded materials. This first exhibition of a new way of living in new circumstances was followed by touring exhibitions that explained the influence of interior design on easier and more rational ways of doing housework and pointed to trends in contemporary design driven by plastics. Questions relating to names of models, brand names and chemical terminology were published in countless brochures

Werbung für Resopalgeschirr, Firma Waca, sechziger Jahre
Advertisement for Resopal tableware, Waca, 1960s

Dieses Jungmädchenzimmer (Entwurf Rolf Grunow) in den neuen Terra-Farben „schiefer", „schilf", „umbra" und in „perlblau" ist in seinen Einzelelementen nach dem Anbauprinzip entwickelt. Durch weitere Einzelstücke kann dieses Zimmer zum Wohn- oder Schlafzimmer ergänzt werden.

Resopalwerbung »Modell: Jungmädchenzimmer«, um 1960
Resopal advertisement "Model: young girl's room," around 1960

dabei auf Trends zeitgemäßer Formgebung durch Kunststoffe hinwiesen. Fragen der Typenbezeichnung, Markennamen und chemische Fachtermini wurden in zahlreichen Broschüren und Informationsschriften mit Hausbuchcharakter veröffentlicht. Diese Ratgeber waren notwendig, um auch die »konservative deutsche Hausfrau«[10], die sich zunächst nur zögernd zum Kauf von Kunststoffwaren entschließen konnte, von den Vorteilen dieses Materials zu überzeugen.

Eine politische Initiative für die »Erziehung zum guten Geschmack« war die 1952 erfolgte Gründung des dem Bundeswirtschaftsminister unterstellten »Rats für Formgebung«. Dieser griff eine Reformidee des Deutschen Werkbunds auf und thematisierte sie neu. Deutsche Erzeugnisse, also auch Kunststoffprodukte, wurden sowohl in Form, Funktion und Fertigung als auch auf ihre Alltagstauglichkeit hin bewertet und prämiert, nicht als staatlich verordnete Geschmacksdoktrin, wie unter dem Naziregime[11], sondern im Interesse der Verbraucher. Die besten Produkte wurden mit dem Prädikat »Gute Form« ausgezeichnet.

Neben dem »Rat für Formgebung« unterstützten etliche dem Deutschen Werkbund zugehörige oder den Ideen des Bauhaus nahe stehende Designer eigenhändig das Bestreben, Industrieerzeugnisse durch hohe Ansprüche an Qualität und Anmutung verkaufsfördernd zu gestalten.[12] Auch die Neue Gemeinschaft der Wohnkultur e.V. (WK), der sich seit dem Ersten Weltkrieg immer mehr Einrichtungshäuser und Handwerksbetriebe angeschlossen hatten, war bemüht, qualitätvolle und angemessene Einrichtungen zur Verfügung zu stellen. Seit 1949 wurden hier die WK-Sozialmöbel (WKS) entwickelt, bei denen in zunehmendem Maße auch die Resopalplatte zum Einsatz kam.[13] All diese Initiativen verhalfen durch ihr Engagement für eine bessere, materialgerechtere Gestaltung den Kunststoffen zum Durchbruch. Strapazierfähigkeit und Funktionalität gepaart mit dem sich neu entwickelnden demokratischen Lebensstil, der übersichtliche Formen mit gesellschaftlichen Freiheiten verband, waren die Kriterien, die insbesondere der Kunststoffplatte den Einzug in die Haushalte erleichterte.

Die neuen Möglichkeiten der Kunststofftechnologie förderten auf der anderen Seite die Ästhetik der neuen Wohnform zunächst im Sinne einer »rechtwinkligen variablen Box«, denn diese ließ sich ganz den Bedürfnis-

and information material that had the character of a home companion. These books of practical advice were needed to convince even the "conservative German housewife,"[10] who was still hesitant about buying plastic goods, of the benefits of this material.

The founding in 1952 of the *Rat für Formgebung* or German Design Council, under the auspices of the Federal Minister of Economic Affairs, was the result of a political campaign on "Education in good taste." This design council took up one of the Deutscher Werkbund's reform ideas and amended it. German goods, which included plastic products, were evaluated for their form, function and quality of manufacture, serviceability in everyday life and distinctions awarded, not in the sense of a government-ordained doctrine on taste such as the Nazis had introduced,[11] but in the interests of the consumer. The best products received the distinction *Gute Form* or Good Design.

As well as the Rat *für Formgebung,* many designers who belonged to the Deutscher Werkbund or had affinities with the ideas of the Bauhaus supported with their own work the attempts to design industrial products in a way that would promote sales by following high standards of quality and appearance.[12] The "Neue Gemeinschaft der Wohnkultur e.V.," a trade association which an increasing number of furniture stores and crafts businesses had joined since the First World War, also made every effort to produce good quality, appropriate furnishings. From 1949, WK-Sozialmöbel was developed in this context, in which Resopal laminated panels were increasingly used.[13] In their commitment to improved design that took better account of the material, all these campaigns helped plastic to make a breakthrough. Its hard-wearing quality and functionality, combined with the new democratic way of life that was evolving in which simple forms were associated with social freedom, were the criteria that enabled plastics to establish a foothold in German homes. That was particularly true of the plastic laminated panel.

On the other hand, the new possibilities offered by plastics technology pushed the new aesthetics of domestic interiors towards the "right-angled variable box," which could be modified to suit the needs of the individual. By manufacturing smooth, seamless boards in different colour combinations with an easy-care surface that had a sparkling sheen

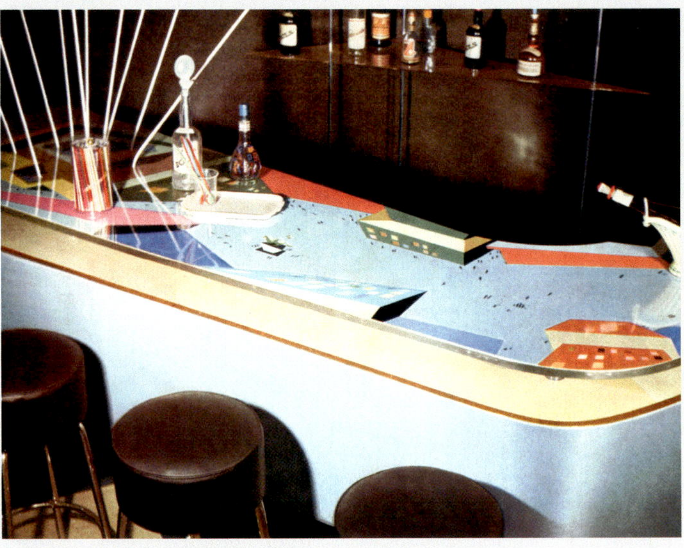

Kinderzimmer, Resopalstand Kunststoffmesse Düsseldorf, 1952
Children's room, Resopal stand at the Düsseldorf plastics show, 1952

Bartresen, Resopalstand Kunststoffmesse Düsseldorf, 1952
Bar counter, Resopal stand at the Düsseldorf plastics show, 1952

sen des Individuums anpassen. Durch die Herstellung von glatten, naht-losen Platten in verschiedenen Farbkombinationen, die durch ihre pfle-geleichte Oberfläche schimmernden Glanz und damit Sauberkeit und Eleganz ausstrahlten, ließen sich Möbel in jeder Form, Größe und für jeden Bedarf produzieren. Basierend auf dem rechtwinkligen Baukasten-system konnte die Möblierung der Wohnung mit steigendem Wohlstand mitwachsen und je nach Geschmack, individuellen Wünschen und finan-ziellen Möglichkeiten zusammengestellt werden. Ein weiteres Argument für die Kunststoffplatte war ihre vielseitige Einsetzbarkeit: als Schrank-wand, Beistelltisch oder Kommode. Es gelang, durch die Kombination von farbigen Flächen und stützenden Elementen in dezent organisch schwingenden Formen – meist bei Sitz- und den so beliebten Kleinmö-beln wie dem Nierentisch – geistige Beweglichkeit und neue Lebensart zu demonstrieren. Diese neue Wohn- und Lebenskultur, die auch helfen sollte, das, was hinter einem lag, zu vergessen, wurde als der erstrebens-werte Stil der neuen Zeit präsentiert. Mit Kunststoffen die Natur in Form und Farbe zu imitieren, wie vor dem Krieg, wurde zunächst vermieden, vielmehr ging es darum, die Andersartigkeit des Materials als Modernis-mus zu betonen. Dabei erhöhten geschwungene Konturen das verkaufs-fördernde Image des Materials und damit den Wert des Produkts.

1952 gelang es der Industrie mit der Kunststoffmesse »K52« in Düssel-dorf, nicht zuletzt dank eines umfangreichen didaktischen Begleitpro-gramms, Kunststoff-Marken und Kunststoff-Erzeugnisse im kollektiven Bewusstsein erstmals stärker zu verankern. Die H. Römmler GmbH aus Groß-Umstadt war mit einem viel beachteten Messestand vertreten, der von Jupp Ernst gestaltet war. Ihm verdankte die Firma eine im besten Sin-ne spielerische und elegante Darbietung der Resopal-Schichtstoffplat-te und ausgewählter Produkte aus Resopal-Pressmasse, dem Werkstoff, den die H. Römmler AG bereits vor 1930 an ihrem ehemaligen Stammsitz in Spremberg entwickelt hatte. Nicht nur die zahlreichen Einsatzmög-lichkeiten der in vielen Farben erhältlichen Platte vom Möbelbelag bis zur Wandverkleidung wurden vorgestellt. Auch im so genannten Unter-druckverfahren hergestellte Resopal-Bilder von Künstlern für die Aus-schmückung eines Kinderzimmers oder als Oberfläche eines Bartresens waren als Neuheit zu bewundern. Dass sich die Resopalplatte auch dem

and therefore radiated cleanliness and elegance, furniture could be pro-duced in any size, any design and to suit any requirement. The right-angled modular system meant that people could furnish their home to suit their pocket and individual taste and then gradually upgrade it as they became more prosperous. Another argument in favour of the laminated panel was its versatility: it could be used for wall units, occasional tables or dressing tables. It was possible to demonstrate intellectual flexibility and a new way of living simply by combining coloured surfaces with support ele-ments in subtly organic curved shapes – usually chairs and sofas or the small pieces of furniture that were so popular at the time, such as the *Nierentisch* or kidney-shaped table. These new trends in lifestyle and home decor, which were also meant to help people forget the immediate past, were presented as the style of the new age that people should aspire to. Using plastics to imitate both the forms and colours of nature, as had been attempted before the War, was avoided; the concern was rather to stress the different nature of this material and link it to the Modernist school of thought. The curved contours that were so popular boosted the mate-rial's sales-promoting image and with it the product's value.

In 1952, the industry staged a trade fair on plastics in Düsseldorf called "K52" and managed – not least thanks to a comprehensive didactic sup-port programme – to anchor plastics brands and products more firmly in public awareness. H. Römmler GmbH from Gross-Umstadt had a high-ly acclaimed stand at the show that had been designed by Jupp Ernst. The company owed to him its presentation of Resopal that was both elegant and playful in the best sense of the word. It included Resopal as a laminate and a selection of Resopal products made of the moulding compound that H. Römmler AG had developed before 1930 at the com-pany's former headquarters in Spremberg. They not only presented the numerous possible uses for the panel that was available in many differ-ent colours, uses ranging from surface finishing for furniture to wall pan-elling. The visitor could also marvel at the Resopal pictures made by art-ists using a so-called vacuum process to decorate a child's bedroom or for the surface of a bar. The fact that the Resopal laminate could also be adapted to make an elegant table with broad sweeping contours was also demonstrated.[14]

Die wachsende Wohnung, Anbaumöbel der Deutschen Werkstätten, 1954
The growing home, modular furniture produced by the Deutsche Werkstätten, 1954

Interieur von Knoll International, 1952
Interior by Knoll International, 1952

eleganten Tisch in großzügig geschwungener Kontur anpassen konnte, wurde ebenfalls demonstriert.[14]

Als unmittelbare Folge dieser erfolgreichen Messepräsentation tauchten in den Einrichtungsempfehlungen der Wohnberatungsstellen vermehrt Kleinmöbel wie Nierentische und Blumentischchen auf, die mit farbigen, maximal 4 mm starken Resopalplatten furniert waren. Sie standen als Symbol des modernen Lebens zwischen den Ohrensesseln mit großkariertem Schonbezug, ihrerseits Symbole der retrospektiven Gesellschaftsvorstellung. Aber die stetig verbesserten Kunststoffe begannen, die Mode in Bezug auf Farb- und Wohnkultur zu prägen. Denn die sich neu formierende industrielle Gesellschaft verlangte auch nach Farbsignaturen, die die Ordnungsprinzipien der teilweise widersprüchlichen Gegenwart spiegelten.

Der Markenname Resopal stellte dabei zu Anfang der fünfziger Jahre noch nicht den Gattungsbegriff dar, in den er sich später verwandelte, als er auch bei Produkten gebraucht wurde, die nicht mit dem hochwertigen Resopal, sondern mit weniger qualitätvollen und billigeren Kunststoffoberflächen ausgestattet waren. Im Gegenteil: Außer, wenn es sich um Küchen handelte, wurde in Fachzeitschriften das monochrome Material, mit dem Tische und Schränke belegt waren, selten näher spezifiziert, sondern nur als Kunststoffplatte bezeichnet. Diese wurde zunächst auch seltener für Wohnmöbel verwandt als zum Beispiel Opalglas und edle Hölzer. Doch die Firma Knoll International, bekannt für ihre vornehmkühlen, rechtwinkligen Einrichtungen, bevorzugte bereits – allerdings nur schwarze und weiße – Resopalplatten als Tischauflage, die dem Ambiente die gewünschte Eleganz des neuen Lebensstils gaben. Die frühen Resopal-Anzeigen hingegen standen unter dem Motto »Die Kunststoffplatte Resopal schmückt mit ihren lichten, frohen Farben jedes Haus und jeden Raum«[15]. Darüber hinaus klärten sie gezielt über die zahlreichen Einsatzmöglichkeiten auch jenseits des Wohnbereichs und über die Handhabung der Schichtstoffplatte auf.

Ein weiterer Meilenstein für die Entwicklung moderner Wohnkultur war der sogenannte »Verbundkreis«, ein 1955 gegründeter Zusammenschluss mehrerer Unternehmen, die sich für die Durchsetzung der »Guten Form« engagierten. Neben der Firma H. Römmler GmbH aus Groß-Umstadt

The direct consequence of this successful trade show was that further small items of furniture such as kidney-shaped tables and plant stands with coloured Resopal veneers to a maximum thickness of 4 mm appeared in the furnishing recommendations of home decor advice centres. They stood as the symbol of modern life between the wing chairs with their large-checked loose covers, which in turn were symbols of a retrospective idea of society. But as plastics constantly improved, they began to dictate fashions in colours and domestic interiors. After all, the newly formed industrial society called for colour signatures that reflected the ordering principles of a present that had some highly contradictory aspects.

In the early 1950s, the brand name Resopal was not yet the generic term it would later become when it was used for products that were made not with this high-calibre product but with cheaper, poorer quality laminates. On the contrary: except in the case of kitchens, trade journals usually referred to the monochrome material used as a surface finish for tables and cupboards merely as a laminate without specifying further. In fact Resopal was not used for living room furniture as often as opal glass or fine wood. But Knoll International, a company known for its coolly stylish, angular furniture had already started to use Resopal laminates as its material of choice for tabletops – albeit only in black or white – to create the elegant atmosphere needed to match the new lifestyle. The early Resopal adverts by contrast ran under the slogan "Resopal laminates with their light and cheerful colours adorn every house and every room."[15] They also focused on providing information about its numerous possible uses beyond domestic interiors and on how to handle the laminated panel.

Another milestone in the development of modern home decor was the so-called "Verbundkreis," an association of several companies founded in 1955, which worked to promote "good design." As well as H. Römmler GmbH from Gross-Umstadt, it included companies such as Max Braun, Radio- und Elektrogeräte from Frankfurt, the Bremen-based Tauwerk-Fabrik, Sisal-Teppiche, Stuttgart-based Knoll International, Möbel und Textilien, G. M. Pfaff Nähmaschinen from Kaiserslautern, porcelain manufacturers Rosenthal, Selb, Württembergische Metallwarenfabrik (WMF)

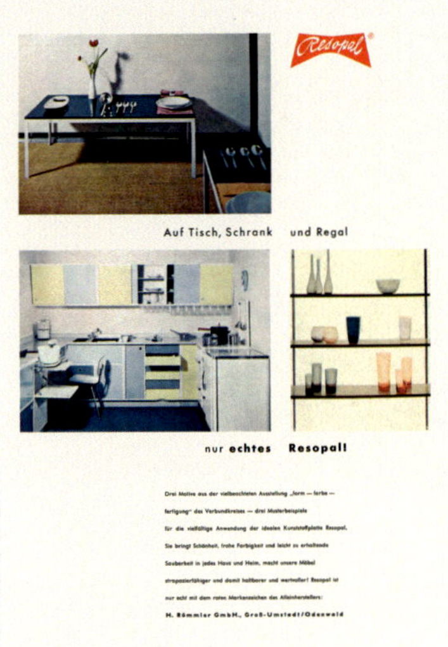

Resopalwerbung mit Fotos aus dem Pavillon des »Verbundkreises«, Bundesgartenschau Dortmund, 1959
Resopal advertisement with photos from the pavilion of the "Verbundkreis," Bundesgartenschau Dortmund, 1959

Resopalwerbung mit Fotos aus der Ausstellung »Form Farbe Fertigung« des »Verbundkreises«, um 1955
Resopal advertisement with photos from the exhibition "Form Farbe Fertigung" organised by the "Verbundkreis," around 1955

gehörten die Firma Max Braun, Radio- und Elektrogeräte aus Frankfurt, die Bremer Tauwerk-Fabrik, Sisal-Teppiche, die Firma Knoll International, Möbel und Textilien aus Stuttgart, das Unternehmen G. M. Pfaff Nähmaschinen aus Kaiserslautern, die Porzellanfabriken von Rosenthal, Selb, die Württembergische Metallwarenfabrik (WMF) und die Firma Gebrüder Rasch Tapeten aus Bramsche zu dieser Vereinigung. Die H. Römmler GmbH war in diesem Zusammenschluss das einzige kunststofferzeugende und -verarbeitende Unternehmen, dessen Schichtstoffplatte ein wesentliches Element zur Produktion für elegantes und zugleich pflegeleichtes neuzeitliches Wohnen auch der übrigen Verbundmitglieder beisteuerte. Ziel dieser Kooperation war, industriell gefertigte Ware in rein funktionaler, schlichter Form von hoher Qualität zu liefern und den Verbraucher dafür zu sensibilisieren. Auf diese Weise hoffte man, wie auch schon der Rat für Formgebung und der Deutsche Werkbund, einen wichtigen Schritt zu einer neuen ästhetischen Formgebung von Massenprodukten zu vollziehen. Einmal mehr wurde dabei betont, dass die Gestaltung einer Ware eine wirtschaftliche Kraft darstellte. Im Verbundkreis stand dabei weniger das jeweilige Unternehmen im Vordergrund als vielmehr die Realisierung der »Guten Form«. Das macht es heute schwer, die einzelnen Beiträge zur neuen Wohnkultur zu differenzieren, war es doch ganz im Sinne des Verbundkreises, leb- und finanzierbare Wohnkultur für alle, ohne einseitige Bevorzugung, zu schaffen. Entsprechend der Satzung sollten die genannten Firmen dafür Sorge tragen, den neuen Wohnstil der Zukunft zu finden. Richtlinie für Produktion und Gestaltung sollte das Kaufverhalten der Kunden sein. Das war eine neue marktwirtschaftliche Zielsetzung, die sich zunächst am noch schmalen Geldbeutel orientierte und den allseits beengten Wohnverhältnissen Rechnung trug. Grazile Regale, Tische und Betten mit den dünnen Querschnitten, Klappmöbel, variable Raumteiler und leichte Stühle mit schmalen Beinen wurden konzipiert. Diese Möbel waren in ihrer Geradlinigkeit eine Sensation.

1958 konnte die Hausfrau in »Kunststoffe und Plasticgeräte für den Haushalt« lesen, dass die »Kunststoffplatte« in der häuslichen Verwendung führend sei. Als die bekanntesten Markennamen wurden Resopal, Ultrapas, Getalit und Formica[16] genannt. Der Handelsname aber, der zum Markenbegriff und damit zum Synonym der Kunststoff-Schichtstoffplatte wur-

and Gebrüder Rasch Tapeten from Bramsche. H. Römmler GmbH was the only company in this association that produced and processed plastics and its laminated panel was an important element in the production of furniture that was both elegant and easy to care for, including goods manufactured by the other members of the group. The aim of this cooperation was to supply high quality industrially produced goods in a simple, functional form and to raise consumer awareness of them. They, like the *Rat für Formgebung* and the *Deutsche Werkbund* before them, hoped to make an important step towards a new aesthetic for the design of mass-produced goods. Once more they emphasised the economic value of product design. The focus in the association was not so much on individual companies as on producing "good design." For that reason, it is difficult today to distinguish the individual contributions to the new home decor; what is certain is that it was the aim of all the member companies to create liveable and affordable home decor for everyone, without elitism. The association's articles of constitution stated that the companies should work towards finding a future for the new style of living. Production and design should take their cue from customers' purchasing behaviour. This was a new market-driven objective that focused initially on people with modest incomes and took account of everyone's cramped living conditions. Lightweight shelving, tables and beds with slender cross-sections, fold-up furniture, variable room dividers and light chairs with slender legs were designed. The simple straight lines of this furniture were sensational.

In 1958 the German housewife read in "Kunststoffe und Plasticgeräte für den Haushalt" that the "laminate" was the leading product in the home. Resopal, Ultrapas, Getalit and Formica[16] were announced as the top brand names. However, the brand that became a generic name and therefore a synonym for plastic laminated board *per se* was Resopal. This laminate was set to give tangible form to the atmosphere of an entire generation's living space. It gave shape to an intellectual and political culture that was asserting itself, albeit only haltingly amongst the middle-classes.

Now an aesthetic and social context had definitively been created that gave a sense of coherence between the style of home decor propagated, its items of furniture and furnishings and actual everyday life. It was only

Resopalanzeigen aus Der Spiegel, 1970
Resopal advertisements from Der Spiegel, 1970

de, hieß Resopal. Diese Kunststoffplatte war auf dem Weg, die Lebensatmosphäre einer ganzen Generation zu verdinglichen. Sie gab einer, sich zugegebenermaßen zunächst nur zögernd in bürgerlichen Kreisen durchsetzenden, geistigen und politischen Kultur Form und Kontur.

Nun war zwischen der propagierten Wohnform, ihren Ausstattungsgegenständen und dem tatsächlich gelebten Alltag endgültig der ästhetisch-gesellschaftliche Kontext geschaffen worden. Erst die Auflösung der historisch gewachsenen Gesellschaftsnormen schuf die Bereitschaft zum Leben mit modernen Zielsetzungen. Die Kunststoffe gaben den Rahmen dazu. Inzwischen war es allerdings auf dem Markt der Wohnkultur und der Kunststoffe zu einer neuen Trennung gekommen. Es gab Firmen, die zeitgemäße, technisch-funktionale Produkte aus Kunststoff produzierten und sich dabei durch eine zurückhaltende Farbgebung auszeichneten. So verwendete die Firma Braun für das Gehäuse ihrer Küchenmaschine, die allein schon in ihrer Funktion den neuen Lebensstil verkörperte, ausschließlich weiße Resopalmasse. Andere Unternehmen entsprachen stärker dem Wunsch breiter Käuferschichten nach einer neuen gemütlichen Wohnkultur und setzten dafür Kunststoffe in bunten Farben, schrillen Mustern und biederen Holzimitaten ein. Die H. Römmler GmbH, die Ende der fünfziger Jahre bereits 60 unterschiedliche Dekore im Programm hatte, bediente beide Kundenströmungen, so dass in mehrfacher Hinsicht das Diktum Gültigkeit erlangte: Kunststoffe bieten etwas für jeden Geschmack und jeden Lebensplan.

Auch in den siebziger Jahren kombinierte die H. Römmler GmbH, die inzwischen den Begriff »Resopal Werk« vor ihren Namen gesetzt hatte, die Vorzüge ihres Produktes wie dauerhafte Qualität, Belastbarkeit und modernes Dekor mit der neuen Lebenseinstellung. So stellte beispielsweise eine ihrer Anzeigen die Verbindung her zwischen der Unverwüstlichkeit des Materials und den neuen Tendenzen antiautoritärer Erziehungsmethoden: »Ihre Kinder kriegen alles kaputt? Irrtum! Resopal ist stärker. Resopal gestaltet die Zukunft«. Lebensräume aus Kunststoff sollten ein Leben ohne Einschränkungen garantieren. Resopal stand nunmehr neben allen übrigen Kunststoffen für dieses neue Lebensbild, in dem zur Individualität erzogen wurde, ohne jedoch gegen soziale Konventionen wie Sauberkeit und Ordnung verstoßen zu wollen. Der deutsche Bundesbürger war endlich im »Kunststoffzeitalter« angekommen.

with the dissolution of the historically evolved social norms that a willingness to live with modern objectives was born. Plastics provided the setting to do just that.

However, in the meantime a new division had occurred in the market between home decor and plastics. There were companies producing contemporary, technically functional plastic products that were characterised by reserved colours. For example, Braun used only white Resopal for the housings of its kitchen appliances, whose very function epitomised the new lifestyle. Other companies met the wishes of broader sections of society for a new kind of homeliness and used plastics in bright colours, loud patterns and conservative wood imitations. H. Römmler GmbH, which at the end of the fifties already had 60 different patterns in its range, served both types of customer, so that the dictum "Plastics offer something for every taste and every life plan" was true in several respects.

In the seventies H. Römmler GmbH, which had prefixed its name with "Resopal Werk," also associated its product's advantages such as durability, resilience and modern décor, with the new philosophy of life. For example, one of its adverts made a link between the indestructibility of the material and the new trends toward anti-authoritarian approaches to child rearing: "Your children will manage to break anything? Wrong! Resopal is stronger. Resopal is the shape of the future." Living spaces furnished in plastic were meant to guarantee a life without restrictions. Resopal and all the other kinds of plastic represented this new philosophy of life, in which children were brought up in a spirit of individuality, without contravening social conventions such as cleanliness and order. The West German citizen had finally arrived in the " plastics age."

Prospekt aus den siebziger Jahren
Brochure from the seventies

RESOPAL IM WOHNHAUS
RESOPAL IN THE HOME

Resopal wurde in der Nachkriegszeit vor allem durch die Möbelindustrie bekannt und hier in erster Linie durch zahlreiche Küchenhersteller, die mit dem Qualitätssiegel Resopal für ihre beschichteten Modelle warben. Unter dem Einfluss von Jupp Ernst entzog sich die H. Römmler GmbH zunächst dem allzu Modischen des Nierentischzeitalters und stellte sich stattdessen im Verbund mit anderen Firmen dem Anspruch der »Guten Form« – dem Credo der Hochschule für Gestaltung in Ulm. Aber jenseits einer Käuferschicht, die der Tradition der Moderne verbunden war, wünschte der zunehmend größere Kundenkreis des Konsumbürgers eine Oberflächengestaltung, die Projektionsfläche sein sollte für seine individuellen Vorstellungen von Schönheit und Ornament. »Diese Leute können wir nicht mit Askese abspeisen« hieß es dazu in einer Resopalschrift. So sah sich die H. Römmler GmbH gezwungen, zumal sie nur ein Halbfabrikat produzierte, Marktinteressen von Kundenfirmen zu entsprechen, obwohl sie die Geschmacklosigkeit der Masse und den puren Geschäftssinn der Möbelhersteller beklagte, »die sich allzu willfährig nach den Wünschen von Lieschen Müller richteten, damit ihre Kassen sich füllten.« Mit ihrer bunten und vielfältigen Angebotspalette stand sie nunmehr im Widerspruch zu den strengen Auffassungen der »Apostel der Moderne«, die für Kunststoffprodukte hauptsächlich die Farben weiß, schwarz oder grau akzeptierten. Vehemente Ablehnung, wiederum nur aus intellektuellen Kreisen, galt auch der Imitation, vornehmlich der von Holz. Römmler konterte: »Der Möbelhersteller kann sich heute nicht mehr auf die Moderne als ein neues Evangelium verlassen. Was zählt, ist die Masse. Unsere industrielle Produktion ist nur realisierbar, wenn wir alles für alle machen.« Die Kritiker befanden: »Resopal darf nicht als Mittel zum Zweck der Vortäuschung eines anderen Materials dienen.« Römmler: »Aber wo kennen wir noch echtes Holz außer bei Sitzmöbeln? Auch das 0,5 mm starke Palisanderfurnier auf einer Holzwerkstoffplatte ist doch nur eine *anerkannte* Illusion.« Zunehmend wurde Resopal in Designkreisen ignoriert. Erst in den achtziger Jahren, im Zeitalter der Postmoderne, verhalf die italienische Gruppe Memphis der dekorativen Schichtstoffplatte wieder zu einem prominenten Platz in der Designgeschichte.

Resopal became famous during the post-war period, primarily through the furniture industry and more particularly kitchen manufacturers who used Resopal as a mark of quality to advertise their laminated products. Under the influence of Jupp Ernst, H. Römmler GmbH initially distanced itself from the all too trendy *"Nierentisch* age" and instead joined with other companies in an association to promote "good design" – the creed of the Hochschule für Gestaltung in Ulm. However, in addition to the type of consumer who felt an allegiance to the traditions of Modernism, a growing circle of buyers were looking for a surface finish that would be a projection screen for their individual ideas of beauty and ornament. "We can't simply fob these people off with asceticism," was the comment in a Resopal publication. Thus, especially since what they manufactured was only a semi-finished product, H. Römmler GmbH felt obliged to meet the market interests of the companies who bought their products, all the while bemoaning the lack of taste of the masses and the fact that furniture manufacturers had nothing but commercial interests at heart, "being all to keen to pander to the wishes of Josephine Bloggs in order to keep their cash registers ringing." With their broad range of brightly coloured products, they were now in conflict with the severe views of the "apostles of Modernism," who accepted the colours white, black or grey for plastics. Imitations were vehemently rejected, especially of wood – but once again only in intellectual circles. Römmler retorted: "Today's furniture manufacturer can no longer rely on Modernism as a new gospel. What counts is the masses. Our industrial production is only feasible if we make everything for everyone." The critics' verdict was: "Resopal must not be used to simulate a different material." Römmler: "But where do you find real wood these days except in chairs? Even the 0.5-mm palisander veneer on a wood-based panel is merely a *recognised* illusion." Resopal was increasingly ignored in design circles. It was not until the eighties, in the age of Post-Modernism, that the Italian Memphis group helped the decorative laminate to regain its prominent place in the history of design.

Dieses Haus

Max Braun nimmt sich als fortschrittliche Firma für Radio- und Fernsehgeräte die international anerkannte Architektur und Raumausstattung zum Vorbild. Deshalb paßt dieser handliche Klein-Super SK 2 für UKW, Mittelwelle und Anschluß für Platten und Tonband gut ins Haus.

Pfaff macht nicht nur ausgezeichnete Nähmaschinen, sondern dazu auch praktische Möbel wie diese. Die Maschine wird in die Platte gesenkt, und man hat einen Schreibtisch, Nr. 9962. Er paßt genau zu WKS-Constructa.

Rasch-Tapeten schaffen — ganz gleich in welchen Räumen — den passenden Hintergrund für klare, gute Möbel. Sie helfen sogar, ungünstig geschnittene Räume optisch zu verbessern. Ihre Muster haben künstlerische Qualität und lassen viele geschmackvolle Einrichtungsmöglichkeiten zu.

Knoll-International, Stuttgart, ist eine der bekanntesten Möbelfirmen der Welt. Alle ihre Entwürfe, von bedeutenden Gestaltern der verschiedensten Länder, zeichnen sich aus durch Ausgewogenheit in Form, Material und Farbe.

hat's wirklich in sich!

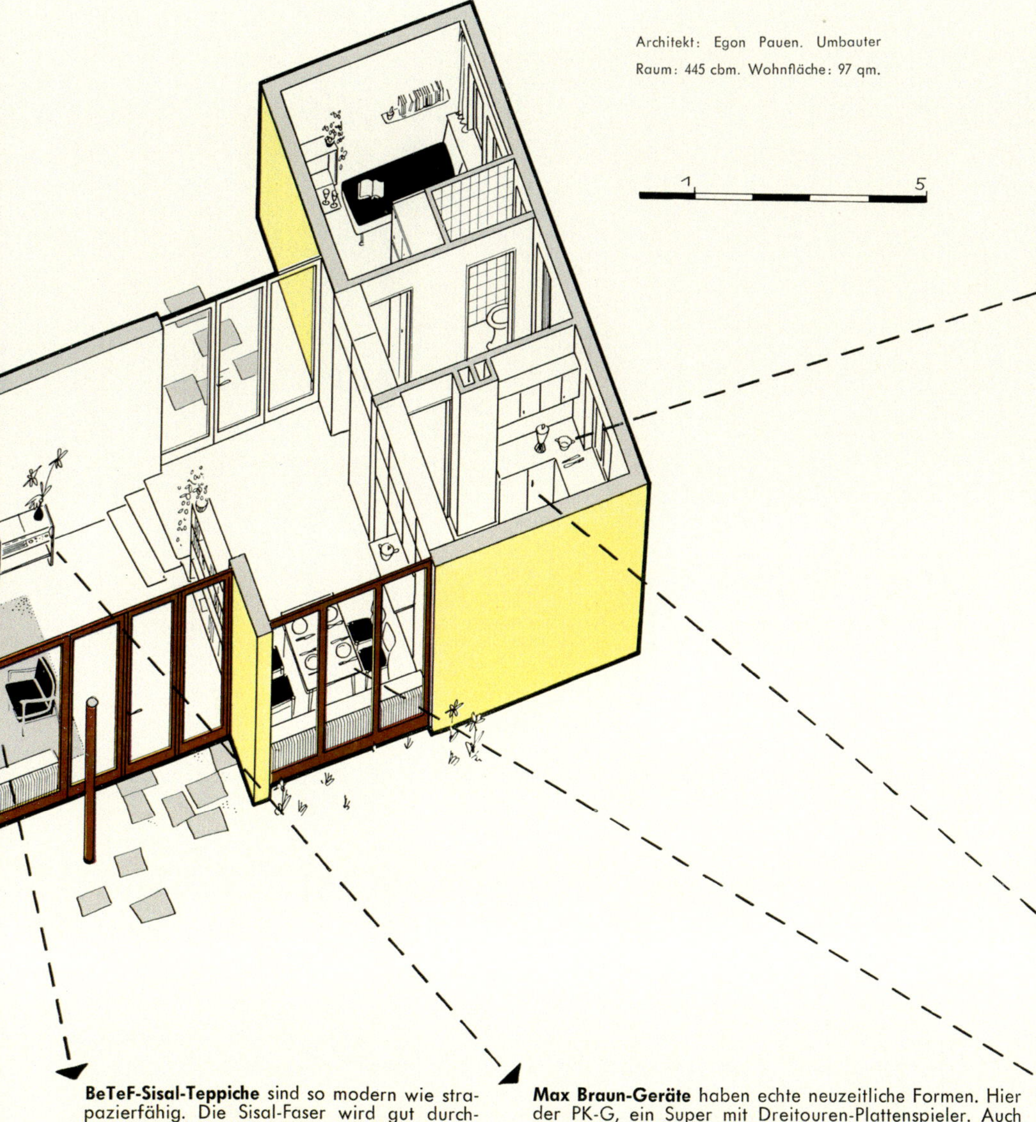

Rosenthal-Porzellan schmückt jeden Raum unseres Hauses, denn es ist zart und fest und von spiegelndem Glanz. Raymond Loewy, einer der bedeutenden Entwerfer für Rosenthal, gestaltete die elegante, gebrauchstüchtige Form „2000", hier mit dem Dekor „Goldfond in Braun".

Resopal heißen die abwaschbaren, robusten Kunststoffplatten der H. Römmler GmbH. Die frohen Farben kommen dem Bestreben entgegen, unsere Küche nicht zum Labor, sondern zum wohnlichen Raum zu machen. Resopal-Platten gibt's im Furnier- und Sperrholzhandel.

BeTeF-Sisal-Teppiche sind so modern wie strapazierfähig. Die Sisal-Faser wird gut durchgefärbt. Die mottensicheren Teppiche sind beidseitig benutzbar, leicht zu reinigen, lichtbeständig, wohnlich und dabei sehr preiswürdig.

Max Braun-Geräte haben echte neuzeitliche Formen. Hier der PK-G, ein Super mit Dreitouren-Plattenspieler. Auch er paßt in jeden hellen, freundlichen Raum. Der Form entspricht die innere Präzision. Ob Sie Schubert hören oder Armstrong — immer ist der Klang rein und unverfälscht.

WMF-Bestecke aus der neuen Produktion der Württembergischen Metallwarenfabrik tun der Hand ebenso wohl wie dem Auge. Das klare, ausgewogene, griffige und dauerhafte Besteck „FORM" 3600 entwarf Professor Wagenfeld.

Saubere Frische . . .

Ein Liebesbrief und eine Rose –

Wo gemixt wird . . .

Die Gärtnerei auf der Fensterbank . . .

Fischgeruch wird weggewischt…

. . . er haftet nicht auf RESOPAL, genausowenig wie Zwiebelduft und andere starke Küchengerüche.

Das ist nur ein Vorzug von vielen, die eine RESOPAL-Küche bietet: RESOPAL verträgt Feuchtigkeit, Fette, Öle, Fruchtsäuren und ist unempfindlich gegen Temperaturen bis zu 130° C.

Zahlreiche neue Farben und Muster erlauben es heute, eine RESOPAL-Küche auch nach individuellen Wünschen einzurichten. RESOPAL ist eben zweckmäßig und schön zugleich.

ist ideal

aber es muß auch wirklich RESOPAL sein
achten Sie immer auf dieses Zeichen

■ **WERBUNG FÜR RESOPAL IN DER WOHNUNG**
Vorige Seite Gemeinsame Anzeige der Firmen des »Verbundkreises«, 1956 **Oben** Anzeigen der H. Römmler GmbH aus den sechziger Jahren **Rechts** Helmut Lortz: Anzeige, die auch in einem Begleitheft zum Katalog der Interbau Berlin 1957 erschien

■ **ADVERTISING FOR RESOPAL IN THE HOME**
Previous page Joint advertisement placed by the companies in the "Verbundkreis," 1956 **Top** Advertisements by H. Römmler GmbH from the sixties **Right** Helmut Lortz: Advertisement that also appeared in a brochure published to accompany the Interbau Berlin 1957 exhibition of built architecture

H. Römmler GmbH. Groß-Umstadt/Odw.

Resopal schmückt

mit seinen lichten, fröhlichen Farben jedes Zimmer. Es spart Zeit, Arbeit und – Nerven. **Resopal** „verwohnt" sich nicht.
Eine Vase wird umgestoßen, eine brennende Zigarette bleibt liegen, ein Blumentopf will Kratzer machen – – –
Kein Grund zur Aufregung. Das Wischtuch nimmt alles fort – und **Resopal** bleibt schön wie am ersten Tag.

Ihre tägliche Freude

Endlich ist ihr Wunschtraum in Erfüllung
gegangen: die RESOPAL-Küche ist jetzt ihre
tägliche Freude.

Alles strahlt in freundlichen Farben.

In einer RESOPAL-Küche macht die Arbeit
soviel Spaß und ist in kurzer Zeit erledigt.

Für Resopal warben nicht nur der Hersteller selbst, sondern auch Küchenfirmen, die das Markenzeichen von Jupp Ernst dabei als Gütesiegel einsetzten. Schon in den fünfziger Jahren wurde der damenhafte Frauentyp mit der Funktionalität der Resopalküche in Verbindung gebracht. In den sechziger Jahren spielten die Qualität des Materials, Eleganz und Luxus eine besondere Rolle, wie die Werbung »Perl« zeigt.

Not only the material's manufacturer advertised Resopal but also kitchen companies who used the logo Jupp Ernst had developed as a quality mark. Even in the fifties, the ladylike type of woman was associated with the functionality of the Resopal kitchen. In the sixties, the quality, elegance and luxurious nature of the material were emphasised, as the "pearl" advertisement illustrates.

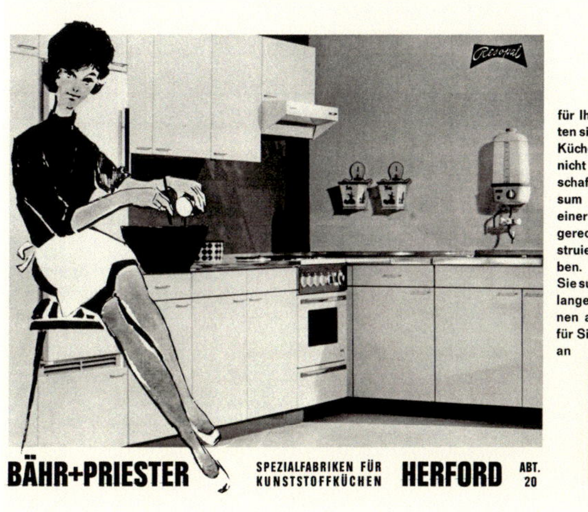

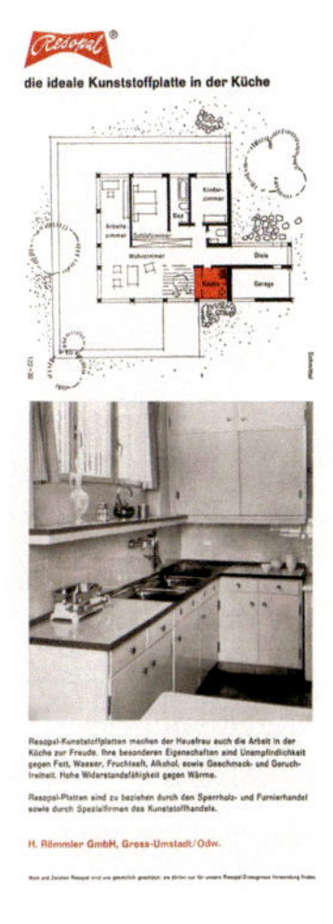

■ **DER EINSATZ DER RESOPALPLATTE IN DER KÜCHE**

1 Formpresstür eines Küchenelements der Firma Gruco aus Massiv-Resopal, 1961 **2** Küche der japanischen Firma TOYO kitchen & living, 2006. Das Dekor „Candy Drops" in Resopal wurde von Lars Contzen entworfen. **3** Küchenschrank aus den fünfziger Jahren **4** AEG-Werbung, sechziger Jahre **5** Modell Segmento SX der Firma Poggenpohl, die Schichtstoffe von Resopal, Duropal (Pfleiderer) und Egger einsetzt, 2006

■ **THE USE OF SHEET RESOPAL IN THE KITCHEN**

1 The moulded door of a kitchen element made by Gruco in solid Resopal, 1961 **2** Kitchen by the Japanese company TOYO kitchen & living, 2006. The "Candy Drops" design in Resopal was created by Lars Contzen. **3** Kitchen cabinet from the 1950s **4** AEG advertising, 1960s **5** Segmento SX model by Poggenpohl, who use Resopal, Duropal (Pfleiderer) and Egger laminates, 2006

1

2

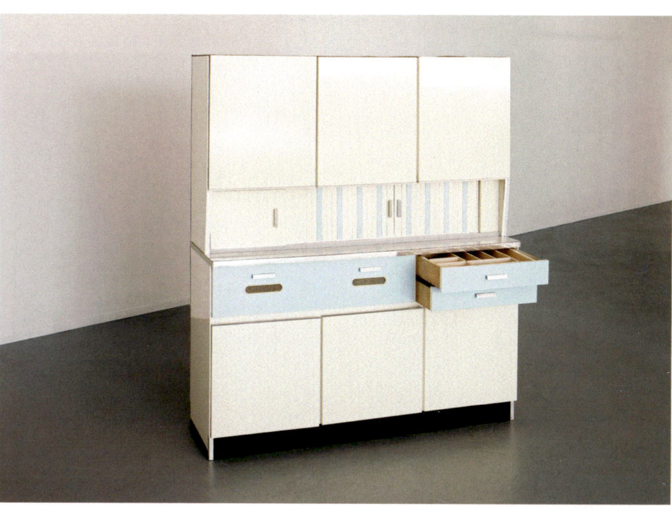

Resopal
Küche
exclusive
diamant
actuel deluxe
mit AEG Einbau-Elektrogeräten
6/1968 Preis- und Typenliste

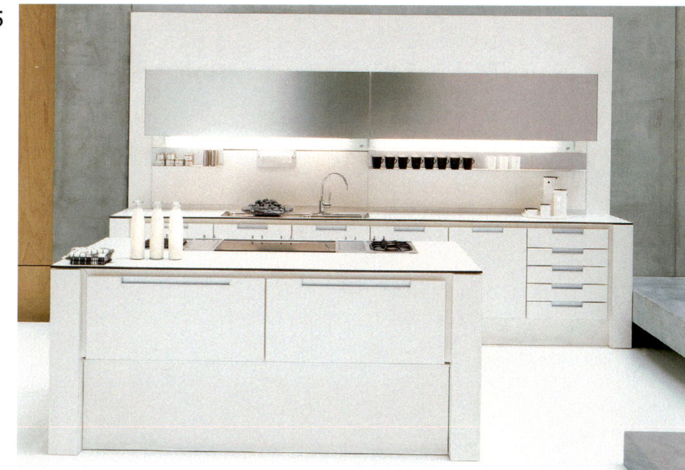

Sicher
Stabil
Zuverlässig
Präzise
Isolierend

Sicherungsautomaten aus
RESOPAL-Pressmassen sind wartungsfrei!
Sie müssen es sein. Deshalb verwendet
man für Sicherungsautomaten
RESOPAL-Pressmassen — zur Sicherheit.
RESOPAL-Pressmassen erfüllen alle
Anforderungen, die an Pressmassen
gestellt werden können. In der Fertigung:
kurze Stehzeiten beim Pressen.
Im täglichen Gebrauch: große
Oberflächenhärte, ausgezeichneter
Oberflächenglanz, keine Rißbildung.
RESOPAL-Pressmassen — so vorzüglich,
wie alles mit dem Zeichen RESOPAL.
Wir stehen Ihnen gern mit genauen
Daten zur Verfügung.
H. Römmler GmbH — 6114 Gross-Umstadt

Dampf
Hitze
Fettdunst
Zugluft

Eine Küchenuhr muß viel ertragen.
Und dennoch soll sie immer hübsch
und sauber aussehen. Deshalb verwendet man dafür RESOPAL-Press-
massen. Denn Uhrengehäuse aus
RESOPAL-Pressmassen sind wider-
standsfähig, antistatisch, leicht zu
pflegen und behalten ihr gutes Aus-
sehen über Jahre.
RESOPAL-Pressmassen erfüllen alle
Anforderungen, die an Pressmassen
gestellt werden können. In der Fertigung:
kurze Stehzeiten beim Pressen.
Im täglichen Gebrauch: große
Oberflächenhärte, ausgezeichneter
Oberflächenglanz, keine Rißbildung.
RESOPAL-Pressmassen — so vorzüglich,
wie alles mit dem Zeichen RESOPAL.
Wir stehen Ihnen gern mit genauen
Daten zur Verfügung.
H. Römmler GmbH — 6114 Gross-Umstadt

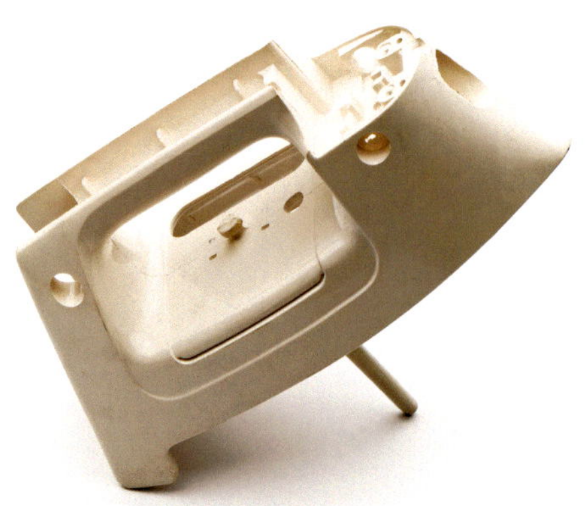

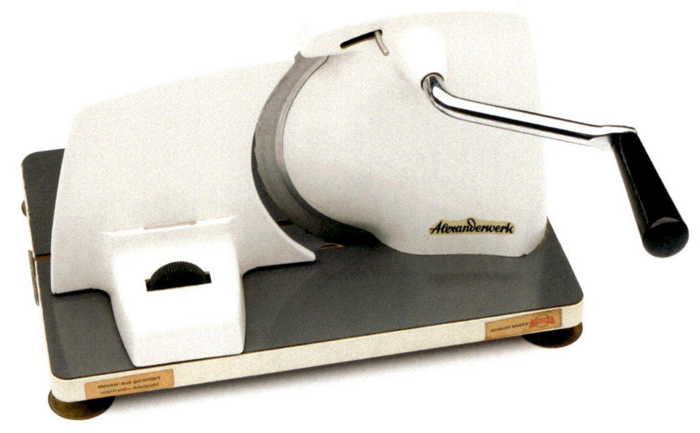

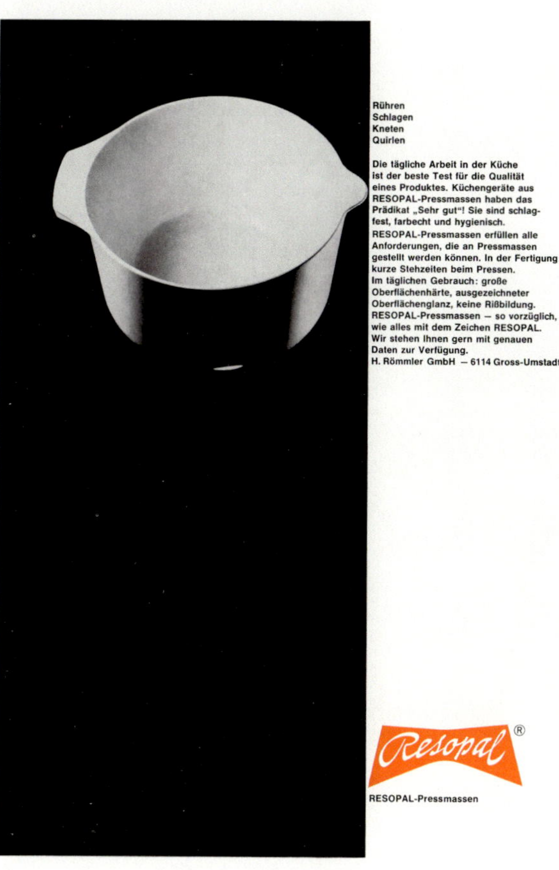

■ RESOPAL-PRESSARTIKEL FÜR DIE KÜCHE

Die Anzeigen aus den sechziger Jahren zeigen Produkte, die zum Beispiel für die Firmen BBC, Junghans und Waca hergestellt wurden. Die
Herstellerfirma von Resopal gehörte von 1938 bis 1987 zur Firma BBC
und produzierte die Gehäuse für deren Haushaltsgeräte wie den abgebildeten »Quirlix«. Unter den aufgeführten Beispielen von Pressteilen
befindet sich auch die Küchenmaschine KM 3/31 der Firma Braun von
1957, die von Gerd Alfred Müller gestaltet wurde.

■ ARTICLES FOR THE KITCHEN IN MOULDED RESOPAL

The advertisements from the sixties show products that were made
for companies such as BBC, Junghans and Waca. From 1938 to 1987 the
company manufacturing Resopal belonged to BBC and produced the
housings for their household appliances such as the "Quirlix" depicted here. The examples of moulded elements listed include the KM 3/31
food processor made by Braun in 1957, which was designed by Gerd
Alfred Müller.

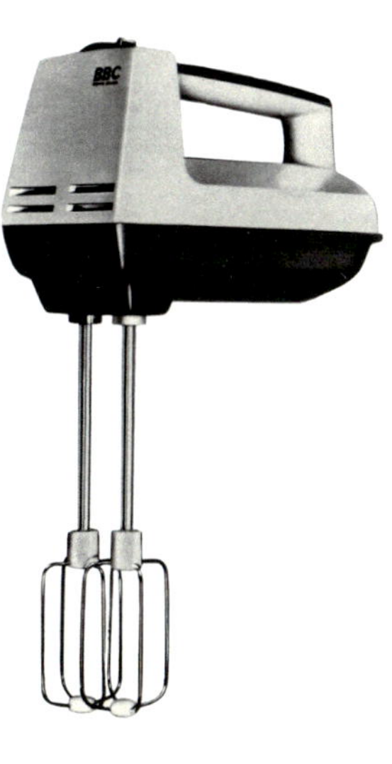

■ **RESOPAL UND NIERENTISCH**

Nach dem Zweiten Weltkrieg setzte sich ein Einrichtungsstil durch, der dem Bedürfnis nach Wohnlichkeit und Luxus mit noch bescheidenen Mitteln nachkam. Wegen der zumeist beengten Wohnverhältnisse waren leichte und vielseitige Möbel gefragt. Die Blumenbank stammt aus der Werkstätte Carl Auböck. Bei der Hausbar sorgt der kleine Einsatztisch über dem Flaschenbehälter für eine einheitliche Fläche. Die folgende Doppelseite zeigt typische Kleinmöbel der fünfziger Jahre.

■ **RESOPAL AND THE NIERENTISCH**

After the Second World War, a style of furnishing became established that used modest means to meet people's need for domestic comfort and luxury. Since living conditions were usually still cramped, light and versatile furniture was needed. The flower bench is from Carl Auböck's workshop. In the bar, the small slot-in table above the bottle holder creates a uniform surface. The following double page shows typical examples of small furniture in the fifties.

■ **RESOPAL IM WOHNZIMMER**

Individuell kombinierbare Anbaumöbel, die viel Stauraum auf kleiner Fläche boten, bestimmten das Wohnen in den fünfziger Jahren, wie die Anzeige von Wolfgang Schmittel deutlich zeigt. Die abgebildete Schrankwand wurde 1967 von Rudolf Lübben entworfen, der hier ostindisches Palisander mit weißem Resopal verarbeitete. Das Mah-Jongg-Spiel, links oben, stammt aus den frühen sechziger Jahren und ist heute ein besonders rares Sammlerstück.

■ **RESOPAL IN THE LIVING ROOM**

Modular furniture that could be combined into individual arrangements and provided a great deal of storage in a small surface area, dominated home decor in the fifties, as the advertisement by Wolfgang Schmittel clearly illustrates. The wall unit depicted was designed in 1967 by Rudolf Lübben, who here combined East Indian palisander with white Resopal. The Mah Jong set, top left, dates back to the early sixties and is now a particularly rare collector's item.

die ideale Kunststoffplatte im Wohnzimmer

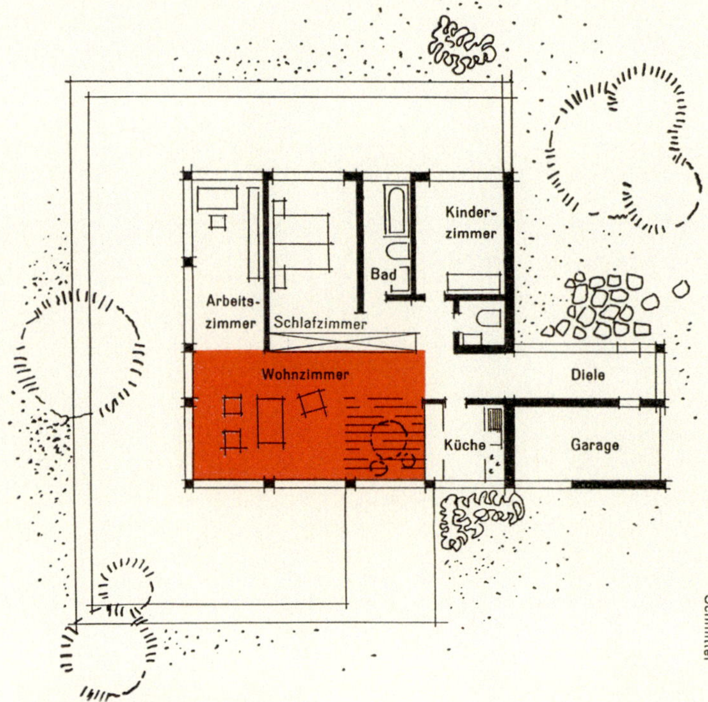

Ein Eßtisch mit Resopal-Platte spart Tischwäsche.
Durch ihre ruhige glatte Oberfläche, Unempfindlichkeit gegen Wärme und
Feuchtigkeit hat Resopal als dauerhafter und schöner Tischbelag in der
modernen Wohnraumgestaltung immer mehr Anwendungen gefunden.

Resopal-Platten sind zu beziehen durch den Sperrholz- und Furnierhandel
sowie durch Spezialfirmen des Kunststoffhandels.

H. Römmler GmbH, Gross-Umstadt / Odw.

■ WEISSES RESOPAL IM WOHNRAUM
Die obere Reihe zeigt Möbel, die von Rudolf Lübben um 1960 bis 1964
entworfen wurden. Bei dem Interieur mit Möbeln der Firma Knoll International handelt es sich um die Musterwohnung im Haus von Alvar
Aalto zur Interbau Berlin 1957. Der Couchtisch von Harry Bertoia ist
mit einer Resopalplatte belegt, die auch in schwarz lieferbar war. Das
Gehäuse des von der Firma Braun hergestellten Plattenspielers »Cockpit« von 1971 besteht aus Pressmasse.

■ WHITE RESOPAL IN DOMESTIC INTERIORS
The top row shows furniture designed by Rudolf Lübben from around
1960 to 1964. The interior with furniture by Knoll International was in
the show flat in the building designed by Alvar Aalto for the Interbau
Berlin, 1957. The coffee table by Harry Bertoia has a Resopal laminate
finish that was also available in black. The housing of the "Cockpit"
record player made by Braun from 1971 is made of moulding compound.

BRAUN cockpit 250

Wenn Kinder nicht schlafen,

sind sie keine Engel. Sie wollen sich in ihrem
Reich austoben dürfen.
Resopal ist geduldig. Es läßt sich bemalen,
schlagen, verkleistern – und wieder abwaschen.
Es schmückt jeden Raum mit seinen heiteren
Farben; und die kleinen Katastrophen im Kinderzimmer können ihm nichts anhaben.
Resopal macht Freude, spart Arbeit und
schont die Nerven der Hausfrau. Schön und
dauerhaft: das ist **Resopal** die Kunststoffplatte
für viele Verwendungszwecke.
Resopal-Platten sind durch den Fachhandel
zu beziehen.

H. Römmler GmbH. Groß-Umstadt/Odw.

■ RESOPAL IM SCHLAFZIMMER

Dargestellt sind eine Anzeige von Helmut Lortz aus den fünfziger
Jahren, eine Schlafzimmerkombination von Rudolf Lübben von 1968
sowie eine Abbildung seines »Kosmosolarium«. Dabei handelt es
sich um den Ausstellungsbeitrag »Wie wohnen und leben wir morgen?« der H. Römmler GmbH zur Internationalen Möbelmesse in Köln
1970. Das automatische Drehbett, über dem eine mobile Lichtkuppel
schwebte, sollte vor allem der Meditation dienen.

■ RESOPAL IN THE BEDROOM

Depicted here are an advertisement by Helmut Lortz from the fifties,
a bedroom suite by Rudolf Lübben from 1968 and his "cosmos solari-
um." This was part of H. Römmler GmbH's stand entitled "Tomorrow's
life and lifestyle?" at the international furniture fair in Cologne in 1970.
The automatic revolving bed, over which a movable dome light hung
hovered, was primarily intended to be used for meditation.

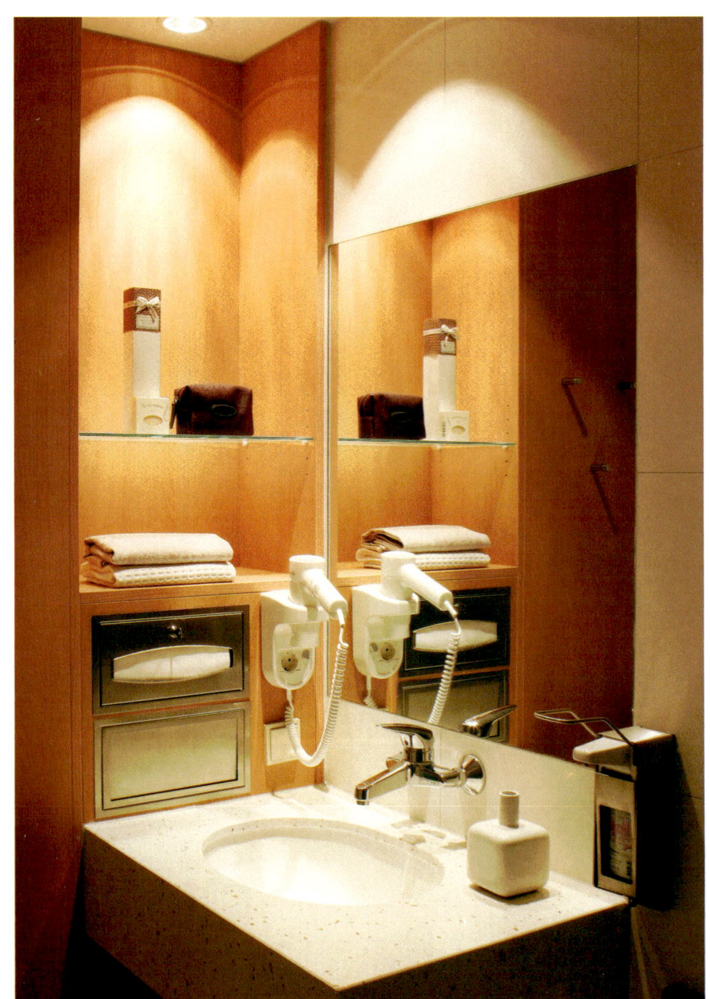

1

2

4

■ **RESOPAL IM BADEZIMMER**

1 Anzeige für den britischen Markt, sechziger Jahre **2** Atelier Gebel: Badezimmer mit Resopal-Wandverkleidung, interdisziplinäre Station, Vivantes Humboldt-Klinikum, Berlin, 2006 **3** Badezimmer mit Resopalmöbeln, siebziger Jahre **4** Seifenschale als Werbegeschenk, siebziger Jahre **5** Verschiedene Pressteile, darunter Gehäuse für Rasierapparate, fünfziger Jahre

■ **RESOPAL IN THE BATHROOM**

1 Advertisement for the British market, 1960s **2** Atelier Gebel: Bathroom with Resopal wall panelling, interdisciplinary ward, Vivantes Humboldt Clinic, Berlin, 2006 **3** Bathroom with Resopal fittings, 1970s **4** Soap dish as an advertising gift, 1970s **5** Diverse moulded parts, including housing for electric shavers, 1950s

3

5

■ **WILSON HOUSE, TEMPLE, TEXAS (USA), 1959**
Der Bau im Stil einer Ranch ist durch einen offenen Grundriss gekennzeichnet. Dargestellt sind der große Wohnraum, der Küchenbereich,
dem sich das Esszimmer anschließt, das Schlafzimmer, das Badezimmer, das aus dem Schlafzimmer führt, und das Gästebad.

■ **WILSON HOUSE, TEMPLE, TEXAS (USA), 1959**
The outstanding feature of this ranch-style building is its open-plan
layout. Depicted are the Great Room, the kitchen with adjacent dining
room, the master bedroom, the bathroom off the master bedroom and
the guest bathroom.

Wilson House, Eingangsseite

Der Gründer des amerikanischen Laminatherstellers Wilson Plastics, Ralph Wilson Senior, nutzte dieses Haus nicht nur als persönlichen Wohnsitz, sondern testete hier auch den Einsatz von HPL- (High Pressure Laminate-) Platten für andere Bereiche als nur Arbeitsflächen in der Küche oder Oberflächen von Tresen. Die meisten Innenwände des Hauses wurden mit dekorativen Schichtstoffplatten verkleidet, auch die Schränke und Einbauten in der Küche und in den zwei Bädern. Im Wohnzimmer wurde eine Wand mit einem geometrischen Muster aus farbigen Platten der Standardkollektion geschmückt. Eine der ersten Ausführungen des sogenannten Postforming zeigen die Küchenarbeitsflächen, bei denen die Schichtstoffplatte nicht an der Kante des Möbels endet, sondern diese umrundet.

Als das mittlerweile in Wilsonart International umbenannte Unternehmen das Haus 1997 von Sunny Wilson, der Witwe des Firmengründers, erwarb, waren die Laminatflächen noch in erstaunlich gutem Zustand. Im folgenden Jahr wurde das Gebäude unter Denkmalschutz gestellt und als Museum für die Öffentlichkeit zugänglich gemacht.

Die Resopal GmbH gehört seit 1998 zu Wilsonart International.

Wilson House, entrance side

Ralph Wilson Senior, the founder of the American laminates manufacturer Wilson Plastics, used this house not only as his home, but also to test the use of HPL (high-pressure laminates) for applications other than just kitchen worktops and counters. Decorative laminates were used to panel most of the house's interior walls, and the kitchen cabinets and fittings in both bathrooms were also made of laminates. One wall in the living room was decorated with a geometric pattern made up of coloured panels from the standard collection. One of the first examples of what is known as postforming is demonstrated in the kitchen worktops, where the end of the laminate is not flush with the cabinet but forms a rounded edge.

When the company, which has since changed its name to Wilsonart International, bought the house in 1997 from Sunny Wilson, the widow of its founder, the laminated surfaces were still in astonishingly good condition. The following year the building was listed and since then has been open to the public as a museum.

Wilsonart International acquired Resopal GmbH in 1998.

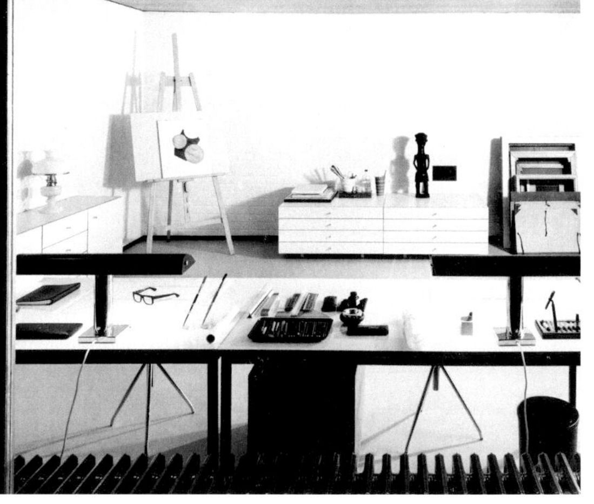

■ HAUS LÜBBEN, MÖNCHENGLADBACH-RHEYDT, 1961

1 Blick ins Wohnzimmer mit anschließendem Esszimmer (und der Küche) **2** Blick in die Nische des Wohnzimmers mit eingebauter Ablage und Barschrank **3** Arbeitszimmer des Hausherrn **4** Einbauküche in Teakholz-Dekor **5** Treppenaufgang mit Einbauten für die Präsentation afrikanischer Kunstgegenstände. Die Türen links sind mit hellgrauem Resopal beschichtet.

■ LÜBBEN HOUSE, MÖNCHENGLADBACH-RHEYDT, 1961

1 View into the living room with adjacent dining room (and kitchen) **2** View into an alcove in the living room with built-in shelves and drinks cabinet **3** Study for the man of the house **4** Fitted kitchen in a teak-design laminate **5** Staircase with built-in units for displaying pieces of African art. The doors on the left are laminated in pale grey Resopal.

4

5

Haus Lübben, Gartenseite

Rudolf Lübben (geb. 1923), bedeutender Möbeldesigner insbesondere für Hoteleinrichtungen und Systemmöbel, war ab 1952 zunächst erster Innenarchitekt und ab 1958 Chefarchitekt des renommierten Einrichtungshauses Kalderoni in Rheydt. »Gegen Resopalplatten mit unifarbenen Oberflächen ohne Muster und Dekor ist überhaupt nichts einzuwenden«, so Lübben. Sein eigenes Haus »ist ein vielfaches Beispiel dafür, wie Resopal zweckmäßig und sinnvoll in allen Räumen anwendbar ist«. Lübben setzte für viele Oberflächen weißes Resopal ein, das er häufig mit edlen Hölzern kombinierte. Bei den Einbauschränken in den Fluren verwendete er hellgraues Resopal. Bestimmend für seine Inneneinrichtungen waren platzsparende, anbaufähige und leicht erscheinende Möbel, die den Raum optimal ausnutzten.

Lübbens Haus wurde im April 1967 in einem ausführlichen Artikel in »Architektur und Wohnform« vorgestellt. Bemerkenswert ist, dass der Werkstoff Resopal dabei nicht erwähnt wurde. Die damalige Einbauküche, die mit Resopal-Teakholz-Dekor beschichtet war, wurde zum Beispiel als Küche aus »Teak mit weißen Kunststoffarbeitsflächen« bezeichnet.

Lübben House, garden side

Rudolf Lübben (born in 1923), a prominent furniture designer specialising in hotel and modular furniture, joined the renowned furniture store Kalderoni in Rheydt as an interior designer in 1952 and became their chief architect in 1958. "There can be no objection at all to Resopal laminates in single colours without patterns or decorative designs," said Lübben. His own house "is in many ways an example of how practical Resopal is and how it can be appropriately used in all rooms." Lübben used white Resopal for many surfaces, often combining it with fine woods. For the fitted cupboards in the hallways he used pale grey Resopal. The main characteristic of his interiors was their space-saving, easily extendable furniture that had a light appearance and made optimum use of the space.

Lübben's house was featured in a lengthy article in "Architektur und Wohnform" in April 1967. It is remarkable that the article makes no mention of Resopal. The fitted kitchen that had a Resopal imitation teak finish was described, for instance, as "teak with white laminated worktops."

■ **NETZWERKARCHITEKTEN: HAUS OLMS, ROSSDORF, 2003**
Neben dem umfassenden Einsatz von Schichtstoffplatten, sind in den Aufnahmen die horizontalen und vertikalen Durchblicke in andere Räume und Ebenen gut zu erkennen, die das Haus gleichfalls auszeichnen.

■ **NETZWERKARCHITEKTEN: OLMS HOUSE, ROSSDORF, 2003**
Apart from the extensive use of laminates the photographs clearly show the horizontal and vertical glimpses through to other rooms and levels that are also one of the outstanding features of this house.

Haus Olms, Gartenseite

Das Haus von Romana und Hans Olms wurde bereits mehrfach ausgezeichnet. Die Jury des »Häuser Award 2006« befand: »Die bewusste
Reduktion von Details und Materialien, verbunden mit einer klaren Lini-
enführung, machen den besonderen Reiz dieser Architektur aus.«
Der rechteckige, zweigeschossige Kubus, der sich durch eine Glasfront
zum Garten und zur Landschaft hin öffnet, wird von zwei Lufträumen
und den festeingebauten Bädern gegliedert. Alle anderen Wände und
Einbauten lassen sich bei Bedarf demontieren oder verändern. Der Innenausbau und die Einrichtung in der beherrschenden Farbe Weiß mit starken Akzenten in Grün und Orange stammen fast komplett von der Haus-
herrin, die Innenarchitektin ist. Alle Schrankwände und Kastenmöbel
wurden aus Schichtstoffplatten gefertigt, unter Verwendung von Resopal und anderer Fabrikate. 2005 erhielt das Einfamilienhaus den Archi-
tekturpreis »Vorbildliche Bauten im Land Hessen«. In der Würdigung
heißt es: »Die Anlage erhält durch die gut überlegten, offenen, ineinan-
der greifenden Raumfolgen und Gestaltungsüberlegungen eine großzügige Wohnqualität.«

Olms House, garden side

The house for Romana and Hans Olms received several awards. In the
opinion of the jury for the "Häuser Award 2006:" "The special appeal of
this architecture is its deliberate reductionist detailing and use of materials, combined with clarity of lines."
The two-storey cuboid building, with its glazed façade opening it up
to the garden and surrounding countryside, is articulated by two full-
height spaces and fixed bathrooms. All the other walls and fixtures can
be dismantled or modified as required. The interiors and furnishings,
predominantly white with strong accents set in green and orange, are
almost entirely the work of Romana Olms, who is an interior designer.
All the wall units and case furniture were manufactured in laminates
using Resopal and other makes. In 2005 this house won the architectural prize "Vorbildliche Bauten im Land Hessen"(Exemplary buildings
in the state of Hesse). The jury's report concludes: "The ensemble has a
spacious quality thanks to its well thought-out, open, intersecting spatial sequences and design details."

Für Freunde · 1956 | Helmut Lortz   Darmstadt-Arheilgen   Jakob-Jung-Straße 12

RESOPAL UND DIE KUNST
RESOPAL AND ART

Für die Kunststoffmesse 1952 in Düsseldorf wurden von Jupp Ernst erstmalig Künstler beauftragt, den Werkstoff Resopal auch in der Kunst einzusetzen. Ernst Oberhoff zum Beispiel schuf damals ein Wandbild für ein Kinderzimmer, das auf dem Firmenstand präsentiert wurde, und gestaltete ein weiteres Bild für das Restaurant der Messe. Üblicherweise wurde bei der Anfertigung der Kunst aus Resopal das Bildmotiv in Tempera- oder Aquarellfarben auf ein Spezialpapier aufgetragen. Dieses wurde dann bei der Herstellung der Resopalplatte eingearbeitet, so dass die Darstellung unter einer schützenden, glasklaren und sehr harten Kunstharzschicht lag. Farbenfrohe Bilder aus Resopal, meist aus mehreren Tafeln zu einer großen Fläche zusammengesetzt, wurden in den nachfolgenden Jahren überaus populär und schmückten zahlreiche öffentliche Orte wie Bahnhöfe, Gaststätten, Krankenhäuser, Universitäten, Schulen und Kindergärten.

Ein besonderes künstlerisches Verfahren entwickelte HAP Grieshaber für seinen Zyklus »Vierzehn Collagen unter Resopal«. Er legte jeweils zwei mit Kunstharz vorbehandelte, farbige Zelluloseschichten übereinander, schnitt die Papiere ein und füllte die Zwischenräume nach Art von Intarsien mit anderen Farbpapieren aus. Dazu verwendete er Dekorpapiere aus dem Sortiment der H. Römmler GmbH.

Geriet die aus Resopal gestaltete Kunst der fünfziger Jahre aufgrund ihres Gebrauchscharakters und wegen ihres Einsatzes als »Kunst am Bau« zunehmend in Misskredit (diese Argumente dienten später deren Demontage und Zerstörung), entstanden autonome Werke von Künstlern, die sich einer kleinbürgerlichen Vereinnahmung vollständig entzogen. Der amerikanische Künstler Richard Artschwager arbeitete seit den sechziger Jahren mit dem »schaurig-schönen Material«, das er »plötzlich zu lieben begann«, weil er den »Anblick der edlen Hölzer satt hatte«. Denn »wenn man es nahm und etwas daraus machte, war es zur gleichen Zeit ein Objekt und ein Bild von etwas«.

Wie die folgenden Beispiele zeigen, eröffnete die künstlerische Auseinandersetzung mit dem industriell gefertigten Material neue Horizonte bei der Wahrnehmung des Elementaren, der Dynamik von Fläche und Raum, der Farben und der mikroskopischen Eigenschaften von Oberflächen.

For the plastics trade show of 1952 in Düsseldorf artists were commissioned for the first time by Jupp Ernst to use Resopal in art. Ernst Oberhoff, for example, created a wall picture for a children's room, which was displayed on the company's stand, and also made a painting for the show's restaurant. The usual method of producing these works of art was to create the painting on special paper using tempera or watercolours. This was then incorporated into the laminate production process, so that the image was protected under a very hard layer of clear synthetic resin. Brightly coloured pictures in Resopal, mostly made up of several panels joined together to form a large area, became very popular in the years that followed and decorated many public places, such as railway stations, restaurants, hospitals, universities, schools and nurseries.

HAP Grieshaber developed a particularly artistic process for his cycle "Fourteen collages under Resopal." He placed two coloured layers of cellulose that had been pre-treated with synthetic resin one on top of the other, made incisions in the paper and then using a kind of intarsia technique filled the spaces created with different coloured papers. He used decorative papers from the range produced by H. Römmler GmbH.

Art made in Resopal in the fifties became increasingly discredited due to its commercial character and its use in Germany's "Kunst am Bau" scheme, a policy of spending a percentage of new building costs on art (these arguments would later be used to justify dismantling and destroying the works of art). Nevertheless, artists who totally resisted being co-opted by the middle classes created a number of autonomous works. The American artist Richard Artschwager began working in the sixties with the "horrendously lovely material, [which he] suddenly began to love [because he was] sick of the sight of fine woods ... [and because] when you take it and make it into something it becomes both an object and an image of something."

As the following examples show, the artistic engagement with this industrially fabricated material opened up new horizons in the perception of the elemental, of colours, of the dynamics between the two-dimensional and three-dimensional, and of the microscopic characteristics of surfaces.

■ **BERND KRIMMEL: WANDFRIES, 1954**

Das Kunstwerk, das der spätere Darmstädter Kulturreferent für das Schuldorf Bergstraße in Seeheim-Jugenheim schuf, schmückte zunächst dessen Kindergarten. Nachdem es dort entfernt wurde, geriet es jahrelang in Vergessenheit. Seit 2002 hängt es nun im Schulrestaurant.

■ **BERND KRIMMEL: FREEZE, 1954**

This work, which the artist, who would later become head of cultural affairs on Darmstadt's city council, created for the Bergstrasse school complex in Seeheim-Jugenheim, originally decorated its kindergarten. After it was removed, it lay forgotten for many years. Since 2002 it has been hanging in the school cafeteria.

◄ **UTE BRINCKMANN-SCHMOLLING: WANDBILD, 1959**

Als die Künstlerin den Auftrag erhielt, für das Infektionshaus der Universitätskinderklinik Giessen ein Wandbild zu schaffen, fand sie in Resopal »das ideale Material« für eine solche Aufgabe. Durch den Beschluss, das Infektionshaus abzureißen, ist der Bestand des Kunstwerks akut gefährdet.

◄ **UTE BRINCKMANN-SCHMOLLING: MURAL, 1959**

When the artist was asked to create a mural for the infectious diseases centre in the children's clinic at Giessen University Hospital, she felt that Resopal was the "ideal material" for this type of commission. The decision to demolish the infectious diseases centre has jeopardised the existence of the work of art.

■ **HELMUT LORTZ: POSTKARTEN FÜR FREUNDE, AB 1952**

Auch die Abbildung auf Seite 134 gehört in die Serie von Neujahrsgrüßen aus Resopal des Künstlers, Designers und späteren Hochschullehrers, der als »Lieblingskonkurrent« von Jupp Ernst auch diverse Anzeigen für die H. Römmler GmbH entwarf.

◄ **HAP GRIESHABER: SCHWARZER KÖNIG, 1956**

Dieses Bild ist eine der vierzehn Collagen unter Resopal, die der Künstler für die Universitäts-Kinderklinik Freiburg im Breisgau schuf.

■ **HELMUT LORTZ: POSTCARDS FOR FRIENDS, FROM 1952**

The illustration on page 134 is also part of a series of New Year's greetings in Resopal by the artist, designer and later university lecturer, who, as Jupp Ernst's "favourite rival" also designed a number of different advertisements for H. Römmler GmbH.

◄ **HAP GRIESHABER: BLACK KING, 1956**

This picture is one of fourteen collages under Resopal that the artist created for the children's clinic at the University Hospital of Freiburg im Breisgau.

Die zweiteilige Arbeit mit spiegeln-
der Oberfläche besteht aus Reso-
palintarsien. Der Hintergrundraster
wird durch das Dekor »Mahagoni-
natur« und »Leinen-weiß« gebildet,
die Figuren sind aus schwarzem
und weißem Resopal.

The two-part piece with a reflective surface consists of intarsia
work in Resopal. The background
grid is created by the patterns
"Natural Mahogany" and "White
Linen;" the figures are in black and
white Resopal.

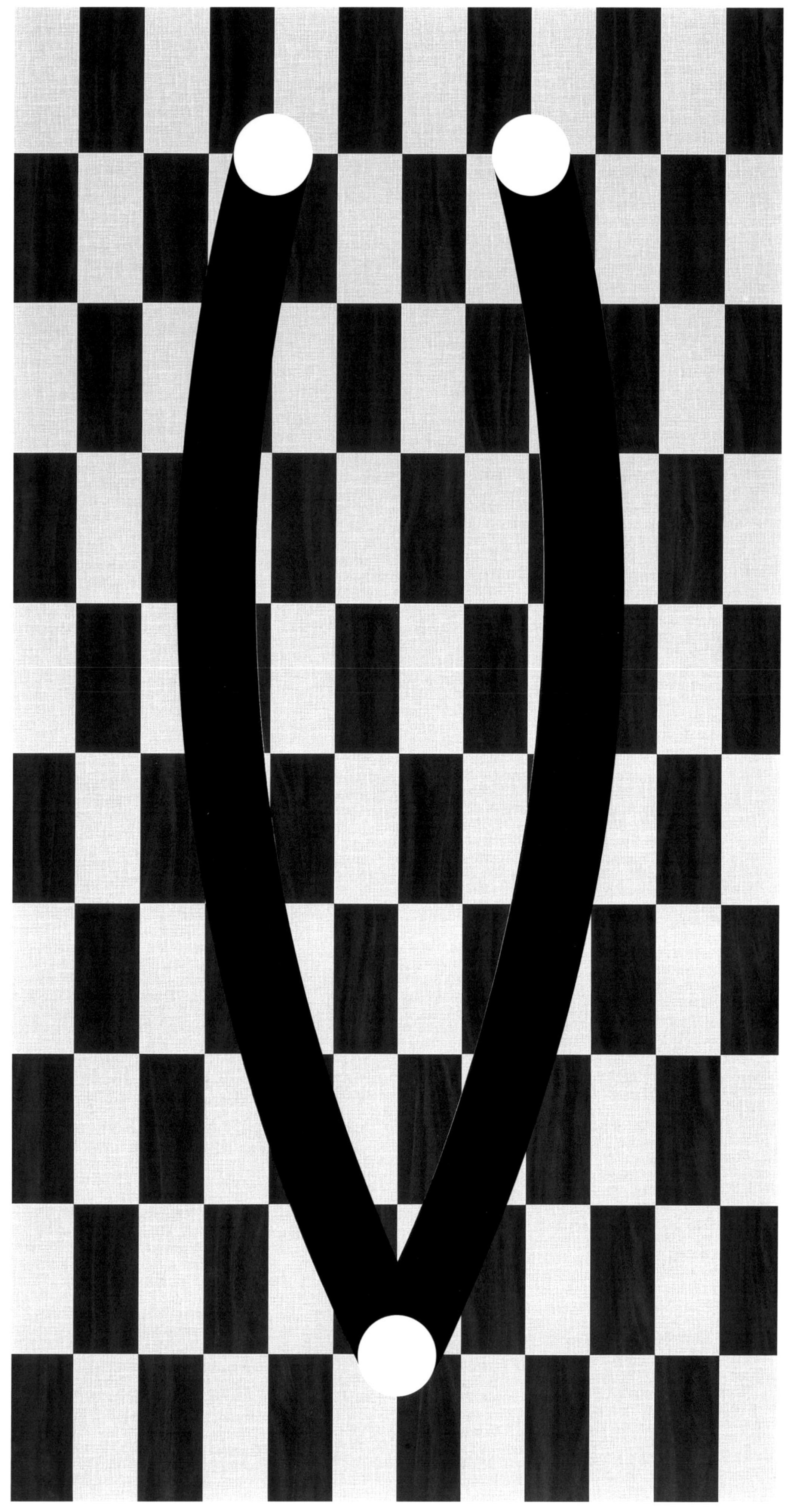

■ **SARAH PELIKAN: DREIKLANG, VIERKLANG, 1993, 4 x 100 x 140 CM**
Die »vier Arbeiten als Serie« sind im Resopal-Unterdruckverfahren entstanden. Die für diese Technik notwendigen Farbpapiere wurden von der Künstlerin entworfen.

■ **SARAH PELIKAN: DREIKLANG, VIERKLANG, 1993, 4 x 100 x 140 CM**
The "four works as a series" were made by using coloured paper designed by the artist pressed between the core sheets and the upper layer of protective resin.

Die Zeichnung, die mit schwarzem Filzstift ausgeführt wurde, befindet sich auf der Resopalplatte. Um das Original zu schützen, ließ es der Künstler mit einer Plexiglasscheibe in einem weißen Holzlackrahmen versehen.

The drawing in black felt pen is on a Resopal panel. In order to protect the original, the artist had it framed in a wooden frame, painted white and covered with Plexiglas.

■ **TOM STARK: RESOPAL 0901 60, ALUCORE WABENKERN, 2005,**
2 x 200 x 125 x 2 CM

Die beiden Arbeiten, C-Print-Verpressungen, gehören zur Gruppe
»Interferenzstrukturen«.

■ TOM STARK: RESOPAL 0901 60, ALUCORE WABENKERN, 2005,
2 x 200 x 125 x 2 CM

The two works, C-print pressings, are part of the "Interference structures" group.

Ulrich Höhns

VORGEHÄNGT UND HINTERLÜFTET. OBERFLÄCHE UND ARCHITEKTUR

CLAD AND REAR-VENTILATED. SURFACE AND ARCHITECTURE

Wer heute ein Haus bauen will, dem steht eine Materialpalette von unge-ahnten Ausmaßen zur Verfügung. Die Vielfalt des Angebots betrifft vor allem das, was von der Architektur sichtbar ist: die Oberflächen. Hierfür hat sich seit den fünfziger Jahren nicht nur das Spektrum traditioneller Baustoffe vergrößert, die ja fortbestehen und dank immer individualisie-render Aufgliederung vergessen lassen, dass es sich dabei um genorm-te Produkte handelt, sondern es kommen seit einigen Jahren in großer Zahl und rascher Folge neue Materialien mit neuen Eigenschaften hin-zu. Einige davon haben sich bereits fest etabliert, andere befinden sich dagegen noch in der Experimentierphase. Sie können fast alles, sie sind fest oder weich, textil oder steinern, monochrom oder vielfarbig, trans-luzent sogar als Beton oder leuchtend, entweder aus sich selbst heraus oder angeregt durch Licht, elektrische Spannung oder chemische Fakto-ren. Es gibt hochfeste Schäume, Nanowerkstoffe und -beschichtungen sowie Gläser, deren Lichtdurchlässigkeit steuerbar ist.

Wer hier den Überblick behalten oder vielleicht überhaupt erst einen Einstieg finden will, kann auf ein wachsendes Netzwerk aus Literatur, Materialdatenbanken und auf die Dienste spezialisierter Materialberater zurückgreifen und findet zwischen *Skins for Buildings: The Architect´s Materials Sample Book* und »Musterkisten« mit Hunderten von origi-nalen Materialproben viel oder bei entsprechend präzisierten Suchkrite-rien auch genau das Richtige. Wer diese Auswahl aber nicht nur als eine Art Rasterfahndung für die optimale technische Gestaltung eines Bau-werks auffasst, sondern dabei auch theoretisch auf sicherem Boden ste-hen und neue Materialien nicht aleatorisch nach Tageslaune, sondern konzeptionell gezielt einsetzen will, der muss besser zurückweichen vor der Materialflut und Grundsätzliches klären.

Der Hinweis auf die »Materialität« eines Hauses oder seiner Kompo-nenten ist ein Modewort in der jüngeren Architektur, oft eine Floskel, die in den Erläuterungsberichten der Architekten fast schon zum Stan-dard beim Versuch der verbalen Verdeutlichung der Entwurfsabsichten gehört. Der Begriff ist nicht eindeutig, sondern ambivalent. Es kann dem einen Entwerfer, der ihn zur Beschreibung einer Idee benutzt, genauso gut darum gehen, Stoffe materialgerecht einzusetzen und diese Stoff-lichkeit mit all ihren Eigenschaften auch »ehrlich« zu zeigen. Ein ande-rer meint damit, dass jedwedes Material, auch das künstliche, natürliche Stoffe imitierende oder verfremdende, seinen technischen Möglichkei-ten entsprechend eingesetzt wird bis hin zur künstlerischen Interven-tion. Hinter der Faszination neuer Möglichkeiten des Materialeinsat-zes in der Architektur scheint auch die Verunsicherung durch, die der damit erklärtermaßen vollzogene Bruch mit Baukonventionen und die stets auch damit einhergehende Umwertung von Materialeigenschaf-ten nicht nur bei den Konsumenten, sondern ebenso bei den Produzen-ten von Architektur auslöst.

Die Dinge, die für einen Bau benötigt werden, waren über lange Zeit-räume und bis weit ins 19. Jahrhundert hinein dieselben: Stein, Holz, gebrannter Ton und Glas. Erst mit dem Beginn der Industrialisierung kamen Metalle hinzu, wodurch sich bisher gültige Maß- und Massen-verhältnisse radikal änderten und sich zunächst technischen Bauten und dann bald auch dem gesamten Feld der Architektur vollkommen neue Möglichkeiten für größere Höhen, Spannweiten, mehr Elastizität und für die Verwirklichung dynamischer Formen von Leichtigkeit und Transparenz eröffneten.

Gottfried Semper hat, ausgelöst durch die Londoner Weltausstellung 1851 mit ihren industriellen, als »stillos« empfundenen Produkten, als einer der ersten Architekturtheoretiker den formbestimmenden Einfluss des Materials auf die Entwicklung eines Stils innerhalb der nützlichen Künste

Anyone wanting to build today has a choice of materials previously undreamt of. The variety on offer is particularly great when it comes to the visible part of architecture: surfaces. On the one hand, The range of traditional building materials available here has expanded since the 1950s and, with increasing customisation, has made it possible to forget that these are actually standardised products. On the other hand, a vast number of new materials with new properties have appeared in rapid succession in recent years. Some of them are now firmly established; oth-ers are still in the experimental phase. They are capable of virtually any-thing: they can be firm or soft, made of textiles or stone, be monochrome or multicoloured, translucent even if they are of concrete, or luminous – either inherently or as a result of interaction with light, electricity or chemical factors. There are highly rigid foams, nano materials and coat-ings and glass that make it possible to control light transmittance.

Anyone wanting to gain a complete overview, or maybe just get a rough idea of what is out there, can make use of a growing network of literature, material databases and the services of specialist materials consultants and, between *Skins for Buildings: The Architect's Materials Sample Book* and "sample boxes" containing hundreds of samples of original materi-als, they will find a great deal to choose from or, if they enter sufficient-ly precise search criteria, they will find exactly what they are looking for. But anyone who sees this selection process as more than a kind of drag-net search for the optimum technical design for a building and would like to stand on firm theoretical ground and use materials not just hap-hazardly guided by the latest whim, but on the basis of a coherent con-cept, would be well advised to take a step back from the flood of availa-ble materials and begin by clarifying a few fundamental points.

Reference to the "materiality" of a building or its components has become a buzzword in recent architecture; it is often nothing more than an emp-ty phrase that has virtually become standard in the design reports writ-ten by architects attempting to explain their design intentions in words. The term is not 100% clear-cut, but rather ambiguous. One designer might use it to describe an attempt to use materials authentically and to show this materiality and all its properties "honestly." Another might just as well believe that any material, including synthetics that imitate natural materials or use them in an unaccustomed way, can be used to the limits of what is technically possible, including any artistic interven-tion. Behind the fascination which the discovery of new uses of materials in architecture holds, lies an insecurity that triggers the declared break with building conventions and the re-evaluation of material properties that goes hand in hand with it – not only on the part of the users but also the producers of architecture.

The elements needed to make a building remained the same over long periods of time, and did not change until well into the 19th century: stone, wood, fired clay and glass. It was only during the industrial revolution that they were joined by iron and metal, which caused radical chang-es in accepted dimensions and proportions and opened up complete-ly new possibilities, initially for engineering constructions and shortly afterwards for the whole of architecture, enabling greater heights and spans, more elasticity and the creation of dynamic forms that were light-weight and transparent.

Gottfried Semper, prompted by London's Great Exhibition of 1851, with its industrial products which he found to be "lacking in style," was one of the first architectural theoreticians to describe how materials deter-mined the development of a style in the applied arts. He triggered an architectural debate that would continue for years to come, in which the idea of "truth to material," probably coined by John Ruskin (*The Sev-*

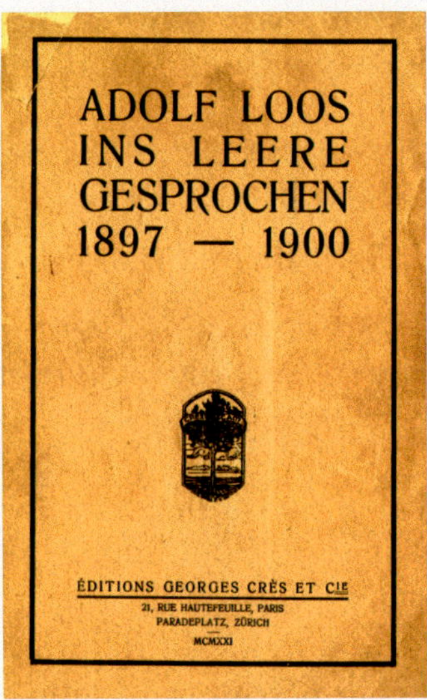

Fabrikgebäude der Factory of Margarete Steiff GmbH, Giengen, 1903

dargelegt und damit eine langanhaltende Architekturdebatte ausgelöst, in der der vermutlich von John Ruskin (*The Seven Lamps of Architecture*) geprägte Begriff der »Materialgerechtigkeit« (»truth to material«) das Aufkommen der modernen Architektur im Industriezeitalter begleitete, deren Wurzeln in die Mitte des 19. Jahrhunderts zurückreichen.[1]

Semper verzichtete in seiner »Bekleidungstheorie« auf Anklagen, wie sie Ruskin gegen die Banalisierung des Bauens in seiner Zeit vorbrachte und dabei auf historische Ableitungen der echten, unverfälschten Baukunst als Grundlage für eine neue setzte. Er wandte sich direkt den veränderten materiellen, stofflichen Grundlagen und ihrer Bedeutung für die gegenwärtige und zukünftige Architektur zu und begründete damit eine heute noch gültige Analogie zwischen Bekleidung, Verkleidung und Architektur. Die Fassade der – bei ihm monumentalen – Bauten löst er dabei von der Tektonik des Hauses ab. Es dauerte mehr als dreißig Jahre, bis auch andere Theoretiker sich der Frage des Materials und seines Einflusses auf die Gestalt der Architektur mit derselben Tiefe annahmen. William Morris, einer der wichtigsten Protagonisten der britischen »Arts and Crafts«-Bewegung, die wiederum entscheidende Impulse für die Gründung des Deutschen Werkbunds lieferte, der sich um die Standardisierung und ästhetische Verbesserung der Industrieprodukte und die Übertragung dieser Prinzipien auf die Entwicklung einer zeitgemäßen Architektur bemühte, äußerte sich 1892 ähnlich über den Einfluss der Baumaterialien auf die Architektur.

Adolf Loos (»Ornament und Verbrechen«) setzte sich 1898 ebenfalls mit den Baumaterialien, ihrem Wert, ihrer Wirkung und ihren Surrogaten auseinander und stellte in provokanter Art im selben Jahr auch das »Gesetz« der Bekleidung in der Architektur auf. »Es muß so gearbeitet werden, daß eine verwechslung des bekleideten materials mit der bekleidung ausgeschlossen ist. Das heißt: holz darf mit jeder farbe angestrichen werden, nur mit einer nicht – der holzfarbe.«[2]

Der erste »ehrliche« Bau der Moderne überhaupt, der dank vollständiger Verglasung seine Stahlkonstruktion und Fassadengliederung offen legt und wegen des Prinzips der Zweischaligkeit mit einer äußeren, vom Boden bis zum Dach durchgehenden und vom Tragsystem abgekoppelten Glasfront die weltweit erste Vorhangfassade trägt, ist das Fabrik-

en Lamps of Architecture), accompanied the emergence of modern architecture in the industrial age, the roots of which go back to the mid-19th century.[1]

In his "dressing theory", Semper avoided accusations of the type Ruskin had levelled against the trivialisation of architecture in his times. Ruskin also advocated using and developing historical forms of authentic, unadulterated architecture as the basis for a new kind of architecture. Semper instead turned his attention directly to the fundamental changes in materials and their significance for contemporary and future architecture. He used this to substantiate the analogy he drew – that still holds true today – between dressing, cladding and architecture. In this analogy he distinguished between the façade of – in his case monumental – buildings and their tectonics. Over thirty years went by before other theoreticians looked in the same depth at the question of materials and their influence on the face of architecture. William Morris – one of the principal protagonists of the British Arts and Crafts movement, wich in turn was a decisive stimulus for the foundation of the Deutscher Werkbund – who was concerned with the standardisation and aesthetic improvement of industrial products and with transferring these principles to the development of an appropriate contemporary architecture, made similar remarks in 1892 about the influence of building materials on architecture.

In 1898, Adolf Loos ("Ornament and Crime") also analysed building materials, their value, their effect and possible surrogates and in the same year proclaimed in his provocative style the "law" of cladding (*Bekleidung*) in architecture. This "law" stated that materials had to be worked in a way that precluded any confusion between the clad material and the cladding. In other words: wood could be painted absolutely any colour except one – wood colour.[2]

Modernism's first ever "honest" building was the factory of Margarete Steiff GmbH in Giengen built in 1903, in which the world-famous soft toys were made. Being glazed in its entirety, it reveals its steel structure and façade articulation and – with its double-skin principle, having an outer glass façade that is completely detached from the structure and runs continuously from the floor to the roof – it has the first curtain wall in the world.

Walter Gropius, Adolf Meyer: Fagus-Werke Fagus Factory, Alfeld/Leine, 1911

Walter Gropius, Adolf Meyer: »Musterfabrik« "Model Factory", Köln Cologne, 1914

Walter Gropius: Bauhaus-Gebäude Bauhaus building, Dessau, 1925

gebäude der Margarete Steiff GmbH in Giengen von 1903, in dem die berühmten Spielzeugtiere hergestellt wurden.

Die deterministische Sachlichkeit dieses von Richard Steiff, dem Neffen der Firmengründerin, maßgeblich mitverantworteten Industriegebäudes ist entwaffnend. Sie spiegelt sich in der zweckgerechten, dienenden Verwendung industrieller Materialien unter Verzicht auf Dekor und Repräsentation wider, die aus sich selbst heraus eine ästhetische Aussage treffen, was von den folgenden, viel bekannteren Einzelobjekten der frühen Moderne in Deutschland nur eingeschränkt behauptet werden kann. Die Fagus-Werke von 1911 in Alfeld von Walter Gropius folgen diesem Prinzip, während bei der von ihm gemeinsam mit Adolf Meyer entworfenen »Musterfabrik« mit Bürogebäude auf der Ausstellung des Deutschen Werkbunds 1914 in Köln bei aller Leichtigkeit der Konstruktion und Feingliedrigkeit der Form ein gewisses Pathos der Monumentalität und der unbedingte Wille zur Darstellung eines Kulturauftrags mitschwang.

Eine *Tour d´Horizon* durch die deutsche Baugeschichte der Moderne des 20. Jahrhunderts mit Blick auf die Fassaden einiger der wichtigsten Häuser und Typen offenbart, dass der radikal- modernistischen Steiff-Lösung bis auf das Dessauer Bauhaus-Gebäude von Walter Gropius aus dem Jahre 1925 nichts Vergleichbares folgte. Zwar lösten viele der Protagonisten des Neuen Bauens bei ihren Häusern die tradierten Bezugssysteme zwischen Innen und Außen auf, aber sie blieben in der Regel doch bei konventionellen Baustoffen. Es kamen allenfalls Brüstungsverkleidungen aus opakem Glas hinzu, oder es wurden hinterleuchtete Gläser eingesetzt, um die Nachtwirkung der Bauten zu steigern. Eine adäquate Umsetzung beispielsweise von Möglichkeiten des Stahlbaus im Häuserbau unterblieb, von wenigen Versuchstypen abgesehen. Die von Le Corbusier favorisierte und auf die Architektur übertragene Maschinenästhetik erzielte er in der Realität der »weißen Kuben im Licht« nur mit bedingter Präzision durch rein handwerkliche Techniken, die als solche jedoch unsichtbar blieben.

Sein gemeinsam mit Pierre Jeanneret entworfenes Doppelhaus in der Stuttgarter Weißenhofsiedlung von 1927, jener ersten international beachteten Mustermesse des Neuen Bauens für den »gebildeten

The deterministic objectivity of this industrial building – which Richard Steiff, the nephew of the company's founder, had a decisive role in designing – is disarming. It is reflected in the functional, subordinate use of industrial materials, devoid of ornament and grandiose symbolism, which make an aesthetic statement purely by virtue of their inherent qualities, something that cannot be maintained without a degree of reservation of later buildings of early Modernism in Germany, some of them far more well-known. The Fagus Factory built in 1911 in Alfeld by Walter Gropius follows this principle, whereas despite its light structure and slender form, the "model factory" with office building designed in conjunction with Adolf Meyer for the 1914 Deutscher Werkbund Exhibition in Cologne had overtones of somewhat sentimental monumentality and the will to display a cultural purpose at all costs.

A *tour d´horizon* of the architectural history of Modernism in the 20th century, focusing on the façades of some of the most important buildings and building types, reveals that with the exception of the Bauhaus building in Dessau, designed by Walter Gropius in 1925, nothing followed that could match the radically Modernist Steiff scheme. Many protagonists of the *Neues Bauen* movement abolished the traditional systems of reference between interior and exterior in their buildings, but they nevertheless usually stuck to conventional building materials. At best they added opaque-glass spandrels or used backlit glass to enhance the effect of the buildings at night. However, there was no adequate translation into practice of the possibilities for using structural steel in buildings, apart from a handful of experimental buildings. The aesthetics of the machine that Le Corbusier favoured and transposed to architecture were actually achieved by him in his "white cubes in light" with limited precision by using pure crafts techniques that remained hidden.

The semi-detached house designed by Le Corbusier in conjunction with Pierre Jeanneret in 1927 on Stuttgart's Weissenhof estate, the first internationally acclaimed model housing exhibition staged by the *Neues Bauen* movement for the "educated middle classes" (Mies van der Rohe), was, like those of neighbouring buildings designed by his colleagues, also the result of craftsmanship. The smooth flawlessness merely feigned machine-based production conditions. Steel, which was a new building

Le Corbusier, Pierre Jeanneret: Doppelhaus, Weißenhofsiedlung Semi-detached house, Weissenhof estate, Stuttgart, 1927

Otto Bartning: Stahlkirche Steel church, Köln Cologne, 1928

Mittelstand« (Mies van der Rohe), war wie die seiner nebenan bauenden Kollegen ebenfalls das Resultat von Handwerkerleistungen, dessen geglättete Makellosigkeit maschinelle Produktionsbedingungen lediglich vorspiegelte. Der zumindest für den Einsatz im Wohnungsbau neue Baustoff Stahl, der die schlank profilierten horizontalen oder vertikalen, in jedem Fall aber ungedämmten Fensterbänder erst erlaubte und sie zu einem wesentlich gestaltbestimmenden Erkennungszeichen der Moderne machte, wurde zu konventionellen, rhythmisch gegliederten Rahmensystemen gefügt, die im Prinzip auch aus Holz hätten bestehen können. Als Träger oder Betonarmierung für größere Spannweiten, größere Auskragungen und schlanke Stützen- und Deckenprofile eingesetzt, blieb das Material bei verputzten, verklinkerten oder mit Naturstein verkleideten Häusern dienend im Hinter- und Untergrund und trat nur sehr selten, dann aber durchaus spektakulär in Erscheinung.

Bei Mies van der Rohes Deutschem Pavillon auf der Weltausstellung 1929 in Barcelona, dem wichtigsten gebauten Manifest der Moderne, trugen acht zierliche, verchromte Stahlstützen (und noch ein paar unsichtbare in den angeblich freistehenden Wänden) das flache Dach. Nebenbei war der Bau mit seiner programmatischen Transparenz ein erstes Beispiel für den erforderlichen Lernprozess des mit dem »fließenden Raum« und seiner gläsernen Begrenzung unerfahrenen Publikums, das »regelmäßig gegen die Scheibe« lief oder in das kleine Bassin im Innenhof fiel, »da schwer zu sehen ist, daß es Wasser ist«.[3]

Otto Bartnings Stahlkirche für die »Pressa«-Ausstellung in Köln 1928 trug ihr architektonisches Programm bereits im Namen, und viele Besucher empfanden es als eine Provokation, dass ein Gotteshaus innen und außen sichtbar aus diesem Material mit seiner technischen Anmutung errichtet worden war.

Es findet sich bei diesen Beispielen aber noch keineswegs ein nicht bestimmungsgemäßer oder seinen Eigenschaften nicht entsprechender Gebrauch des Materials. Vorläufig wird es nur aus seinem technisch angestammten und ästhetisch vertrauten Kontext herausgelöst und in veredelter Form in einen luxuriösen Empfangspavillon verpflanzt oder bestimmt als technischer Werkstoff durch Reihung, Raster und Reduktion formal einen sakralen Raum. Der in den zwanziger und dreißiger Jah-

material, at least for houses, the material which made the slender-profiled horizontal or vertical but always uninsulated ribbon windows possible in the first place and made them into one of Modernism's main identifying design features, was incorporated into conventional, rhythmically articulated frame systems that in principle could have been made of wood. When used as girders or to reinforce concrete, for wider spans, deeper cantilevers and slender column and floor profiles, the material remained in a subordinate role either underground or in the background of rendered, brick- or stone-faced houses. It emerged only rarely, but when it did it usually made a spectacular appearance.

In Mies van der Rohe's German Pavilion at the World Exposition of 1929 in Barcelona, Modernism's most important built manifesto, the flat roof was supported on eight delicate, chrome-plated steel columns (and some invisible columns in the purportedly free-standing internal walls). Incidentally, the building, with its programmatic transparency, was one of the first examples of the learning process that the inexperienced public had to go through with "free-flowing space" and its glazed enclosure; they " regularly [walked into] the glass pane" or fell into the small pool in the courtyard, "because it was hard to see that it was water."[3]

The architectural programme of Otto Bartning's Steel Church for the Pressa exhibition in Cologne in 1928 was evident even in the name, and many visitors found it provocative that this material with its engineering aspirations should be visible on the inside and outside of a house of God.

But in these examples there is no trace of the material being used for any unintended purpose nor in a way that did not match its properties. For the time being, it has merely been removed from its technical origins and aesthetically familiar context and transplanted to a luxurious reception pavilion or used as an engineering material arranged in rows or grids to determine the sometimes reductionist form of a sacred space. The façade's loss of meaning within the hierarchy of values of architectural elements that consequently occurred in the 1920s and 1930s is unmistakeable. Formerly it had portrayed– either legibly or in coded form – the interior of the building and translated it into a particular language, whereas now it has become a more or less autonomous part of the design brief.

In response to the reproach that his buildings were nothing more than

Fritz Höger: Chilehaus, Hamburg, 1924

Hans Poelzig: Verwaltungsgebäude der Office building for IG Farben, Frankfurt/Main, 1932

ren damit einhergehende Bedeutungsverlust der Fassade innerhalb der architektonischen Werteordnung ist unübersehbar. Sie verkörpert nicht mehr – ablesbar oder verschlüsselt – das Innere des Gebäudes und übersetzt dies in eine Sprache, sondern sie wird zu einem mehr oder minder autonomen Teil des Entwurfsprogramms.

Bis auf wenige Ausnahmen, galt für das meist konservative Materialverständnis der gesamten »klassischen« Moderne die Antwort des modernistischen französischen Architekten Robert Mallet-Stevens, der sich explizit mit dem Dekor der Moderne und mit Filmbauten befasst hat. Auf den Vorwurf, seine Bauten seien nichts anderes als dekorative Kunst, entgegnete er: »L´apparance ne trompe pas« – Der Schein trügt nicht.[4]

Es ist erstaunlich, dass es der modernen Bewegung – denn nur sie hätte dazu in der Lage sein können – bis gegen Ende der vierziger Jahre nicht gelang, neue Materialien auch als sichtbare konstruktive und gestalterische Elemente einzusetzen. Was sich dagegen änderte, war der Umgang mit dem vorhandenen Material, seine Aufwertung und Umdeutung vor allem bei Großbauten. Fritz Högers Hamburger Chilehaus von 1922/24, die vielleicht wichtigste Landmarke des Expressionismus, war mit dunklen blaurot-bunten, gebrannten Klinkern dritter Wahl lediglich verkleidet. Der Bau mit seiner markanten, an einen Schiffsbug erinnernden Spitze ist jedoch kein archaischer Monolith, sondern ein künstlerisch ummanteltes modernes Haus mit einer ornamentierten Vorsatzschale, die von einer Stahlbetonkonstruktion gehalten wird, welche wiederum flexible Grundrissaufteilungen ermöglicht. Es geht dabei allein um den Ausdruck, um die Anmutung des Gebäudes, hervorgerufen durch seine Fassade. Dieselbe Technik machen sich natürlich auch Architekten zunutze, wenn sie ihre Bauten mit Natursteinplatten bekleiden, die nicht mehr als sich selbst tragen. Das wichtigste Beispiel dafür ist der mit Travertin verkleidete Stahlskelettbau des Verwaltungsgebäudes der IG Farben in Frankfurt am Main, den Hans Poelzig 1928/32 realisierte. Auch bei diesem Haus löst sich die Wahrnehmung tradierter Materialien von den modernen Konstruktionen ab. Die in ihrem Kern allein aus ökonomischen Gründen zwangsläufig modernen Bauten können jetzt jedes Kleid tragen.

Dasselbe Prinzip gilt für nahezu alle Repräsentationsbauten des »Dritten Reichs«, und heute ist es eine Selbstverständlichkeit, dass Natursteine

decorative art, the French architect with Modernist leanings Robert Mallet-Stevens, who explicitly worked with Modernist decor and with film sets, responded "L´apparance ne trompe pas" – appearance is not deceptive.[4] Apart from a few exceptions, that was true of the usually conservative understanding of materials of all the "Classic" Modernists.

It is astonishing that, by the end of the 1940s, the Modern Movement – after only no-one else would have being capable of doing it – did not succeed in using new materials as visible structural and design elements. However, what did change was their treatment of existing materials, their re-valuation and reinterpretation, particularly in large buildings. Fritz Höger's Chilehaus in Hamburg, dating from 1922/24, which is perhaps one of the most important landmarks of Expressionism, was simply faced in brick of a speckled dark-reddish/blue colour. But this building, with its striking tip reminiscent of the bow of a ship, is not an archaic monolith, but a Modernist building with an artistic cloak in the form of a decorated shell fixed to a reinforced concrete structure, permitting flexible floor plans. The most important thing about it is its expression, the appearance that is produced by its façade.

Architects, of course, use the same technique when they clad their buildings in slabs of natural stone that carry nothing other than their own load. The most important example of this is the office building for IG Farben in Frankfurt am Main, the steel-frame building faced in travertine, which was designed by Hans Poelzig in 1928/32. In this building too the perception of traditional materials is detached from the modern structure. The Modernist buildings that in essence were Modernist purely for economic reasons were now able to wear any clothing.

The same principle applies to virtually all the monumental official buildings of the "Third Reich," and today it is completely taken for granted that natural stone is cut into thin slices and "true to the material" fixed onto façades in such a way that from the joint pattern to the quoins an impression of solidity, traditionalism and credibility is conveyed. Oswald Mathias Ungers´ 1995 Gallery of Contemporary Art at the Hamburg Kunsthalle, a building of light, cream-coloured Portuguese limestone, rising up on a voluminous plinth of dark reddish-brown Swedish granite, is a particularly pretentious, monumental example of this.[5]

Oswald Mathias Ungers: »Galerie der Gegenwart« der Hamburger Kunsthalle
Gallery of Contemporary Art at the Hamburg Kunsthalle, 1995

Egon Eiermann: Kaiser-Wilhelm-Gedächtniskirche Kaiser-Wilhelm Memorial Church,
Berlin, 1963

in dünne Scheiben geschnitten und »materialgerecht« vom Fugenbild bis zu den Ecklösungen an die Fassade gebracht werden, um Massivität, Traditionalität und Glaubwürdigkeit zu vermitteln. Oswald Mathias Ungers' 1995 fertiggestellte »Galerie der Gegenwart« der Hamburger Kunsthalle aus hellem, crèmefarbenen portugiesischen Kalkstein, die sich auf einem voluminösen Sockel aus dunklem, rotbraunen schwedischen Granit erhebt, ist dafür ein besonders prätentiöses, monumentales Beispiel.[5]

Neue Impulse erhielt der architektonische Diskurs um die Ästhetik der Oberflächen in den fünfziger Jahren, als erstmals tatsächlich neue Materialien in größerem Umfang eingeführt oder bereits vertraute aus ihrem angestammten Kontext herausgelöst und in andere Zusammenhänge gestellt wurden. Egon Eiermann ist einer der wichtigen Neuerer auf diesem Gebiet. Er experimentierte mit Wellzement-Platten in »Naturgrau« wie bei einer Fabrik in Blumberg von 1949/51, einem durch formale Reduktion gestalterisch veredelten Industriebau.[6] Für Wohn- wie für Geschäftshäuser führte er ein System komplex in der Tiefe gestaffelter und dadurch besonders leicht wirkender Fassaden ein, deren äußere »Schicht« wie beim Hochhaus für die Abgeordneten des Deutschen Bundestages in Bonn von 1965/69 flatternde, farbenfrohe Textilrollos oder -markisen als Sonnenschutz sind. Neu war die zweischalige farbige Verglasung von Betonsteinen für den freistehenden Turm und das neue Oktogon der Kaiser-Wilhelm-Gedächtniskirche in Berlin von 1956/63, die er zusammen mit dem Glaskünstler Gabriel Loire entwickelte. Neu auch im Material war die von ihm für das Stuttgarter Kaufhaus Merkur 1959/61 entworfene Wabenfassade aus elementierten Keramikformsteinen mit zwei sich kreuzenden, einmal nach innen und einmal nach außen gewölbten Flächen. Diese in der Nahsicht organische, außerordentlich plastische Einzelform verleiht dem Kubus durch ihre rhythmische, jeweils um 90 Grad gegenüber der Nachbarform versetze Anordnung eine fast textil anmutende und dennoch harte Kontur. Andere Architekten entwarfen ähnliche Muster auch aus Hartkunststoffen für Kaufhäuser in Ost und West. Wie ein homogenes Netz überziehen diese Waben die großen Bauten, deren Formen und Funktionen sie anonymisieren, um sie zugleich als Marke zu individualisieren. Diese neuen Oberflächen sind autonome, sich

The architectural discourse about the aesthetics of surfaces received new impetus in the 1950s when modern materials were introduced on a larger scale or familiar materials were removed from their original context and placed in different frameworks. Egon Eiermann was one of the important innovators in this field. He experimented with corrugated cement sheets in "natural grey," in his factory in Blumberg dating from 1949/51, for example, using formal minimalism to give the industrial building a more refined look.[6] For residential and commercial buildings, he introduced a complex façade system with recessed elements, which gave the buildings a very light appearance. Their outermost "layer" consisted of brightly coloured fluttering, fabric blinds or awnings as sunshading, as in the high-rise for the Members of the German Parliament, or *Bundestag*, in Bonn (1965/69). An innovative use of material was to set a double layer of blue glazing in pre-cast concrete blocks in the new freestanding tower and octagon, which he added to the bomb-damaged Kaiser-Wilhelm Memorial Church in Berlin 1956/63. These components were developed in conjunction with glass artist Gabriel Loire. Another innovation in materials was the honeycomb façade he designed for the Merkur department store (1959/61) in Stuttgart, which was made of prefabricated moulded ceramic blocks with two intersecting curved surfaces – one convex, the other concave. Seen close up, these individual forms look organic and extremely plastic; they are in a rhythmic arrangement, each one at a 90-degree angle to the next, giving the cube an outline that looked almost as if it was of fabric and yet hard at the same time. Other architects designed similar patterns, also in hard plastics, for department stores in East and West Germany. These honeycombs covered the large buildings like a homogeneous net, making both the form and function of the building anonymous in order to give them a single identity as a brand. These new surfaces are autonomous, self-explanatory systems without any logical connection with the interior life of the buildings, whose only contact to the outer world is through their ground floor display windows. They orchestrate an optical conundrum of physicality and apparent dematerialisation, an effect produced by way their elements are reflected and perforated: the perfect camouflage for what is essentially a hulking great windowless megaform right in the middle of the city.

Hans Scharoun: Philharmonie, Berlin, 1963

Herzog & de Meuron: „Elbphilharmonie", Hamburg, in Planung
planning stage

selbst erklärende Systeme ohne logische Verbindung mit dem Innenleben der Häuser, deren einziger Kontakt zur Außenwelt ihre Schaufensterfronten im Erdgeschoss sind. Sie inszenieren ein optisches Verwirrspiel aus Körperhaftigkeit und dem durch die Reflexion und Perforation der Elemente hervorgerufenen Anschein ihrer Entmaterialisierung, die perfekte Camouflage einer im Kern ungeschlachten, fensterlosen Großform in der Innenstadt.

Genau das Gegenteil, nämlich die Erzeugung von Präsenz, Massivität und Hermetik strebte Hans Scharouns Berliner Philharmonie von 1956/63 an, die ihre goldschimmernde Metallfassade allerdings erst 1978 erhielt. Die besonders einprägsame bewegte, aufschwingende Trauflinie findet sich in ähnlicher Anmutung, aber größeren Abmessungen bei den gläsern spiegelnden Ansichten der in Planung befindlichen Hamburger »Elbphilharmonie« der Architekten Herzog und de Meuron wieder, die sie auf den monolithischen »Kaispeicher A« (1962/66) von Werner Kallmorgen stellen. »Sie verbindet Backstein-Klassik mit dem kühnen Schwung der Glasfassaden und der Dachlandschaft.«[7] Der Entwurf entwickelt seine suggestive Kraft aus dem harten Kontrast der beiden grundverschiedenen Oberflächen, dem unvermittelten Übereinander von Leichtigkeit und Schwere.

Die Geschichte der aus den unterschiedlichsten Gründen metallverkleideten Bauten reicht von der Bewältigung technisch anspruchsloser Aufgaben bis hin zum 1998 fertiggestellten Jüdischen Museum in Berlin von Daniel Libeskind und darüber hinaus. Virtuos gestaltete Metalloberflächen jeder Farbe, Struktur und Anordnung der Bleche zeichnen heute viele ambitioniert gestaltete Bauten vom Wohnhaus bis zum Museum aus, aber es ist nicht abzusehen, ob diese Oberflächen eines Tages auch als gleichberechtigte Elemente der Architektur akzeptiert werden. Noch besitzen sie Außenseiterstatus, und die Wahrnehmung der Gebäude bewegt sich zwischen dem gesteigerten Interesse für einen ganz oder teilweise verborgenen Inhalt und Verständnislosigkeit angesichts der Wahl eines technisch anmutenden Baustoffs.

Werden diese Bauten gern als »Blechkisten« abqualifiziert, so genießen »Glaspaläste« eine bessere Reputation. Dies ist einer der Erfolge der Moderne, die Glas wie keinen anderen Baustoff von seiner expressi-

Hans Scharoun's Berlin Philharmonie (1956/63), which did not receive its shimmering gold metallic façade until 1978, was aiming for the exact opposite, namely to create a sense of presence, solidity and hermetic separateness. Its particularly striking eaves lines that curve in an upward sweep are echoed, albeit in larger dimensions, in the reflective glass elevations of the "Elbphilharmonie" in Hamburg currently being designed by Herzog and de Meuron, which will sit on the monolithic warehouse "Kaispeicher A" (1962/66) designed by Werner Kallmorgen. "It marries brick Classicism with the bold curves of its glass façades and roofscape."[7] The design's suggestive power stems from the stark contrast between the two fundamentally different surfaces, the unmitigated juxtaposition of lightness and heaviness.

The history of buildings that are clad in metal includes all kinds of different motives ranging from merely solving technically undemanding briefs through to the Jewish Museum in Berlin by Daniel Libeskind completed in 1998. Many ambitious buildings today – from private houses through to museums – are characterised by metallic surfaces designed with great virtuosity, displaying every colour, structure and arrangement of metal sheets, but it is impossible to predict whether these surfaces will one day be recognised as fully fledged architectural elements. They still have outsider status and are viewed either with heightened interest in their completely or partially hidden content, or with complete lack of understanding as to how anyone could choose a material that looked as if it belonged on an engineering structure.

While these buildings are often dismissed as "metal boxes," "glass palaces" enjoy a somewhat better reputation. This is one of the Modern Movement's success stories: Modernist architects worked with glass as with no other material, using it equally well to define or dissolve the shape of a building. Examples range from expressive hyperbole transparent buildings, such as Bruno Taut's pavilion at the Werkbund's 1914 exhibition in Cologne, through to its conceptual successor, the "Headquarters for the glass industry" in Dusseldorf designed by Bernhard Pfau (1951) – the building was subsequently remodelled beyond recognition but was originally, in keeping with its owner's business, completely in glass – through to technically sophisticated structures that were entirely in glass such as

Daniel Libeskind: Jüdisches Museum Jewish Museum, Berlin, 1998

Bernhard Pfau: »Haus der Glasindustrie« "Headquarters
for the glass industry", Düsseldorf, 1951

Werner Sobek: Wohnhaus R 128 House R 128,
Stuttgart, 2000

ven Überhöhung in transparenten Bauten wie Bruno Tauts Pavillon auf
der Kölner Werkbundausstellung 1914 über seinen ideellen Nachfolger,
das konsequent auf den Baustoff Glas abgestellte »Haus der Glasindustrie« in Düsseldorf von Bernhard Pfau aus dem Jahre 1951, das später
zur Unkenntlichkeit überformt wurde, bis hin zu technisch ausgereiz-
ten Ganzglaskonstruktionen wie Werner Sobeks Stuttgarter Wohnhaus
R 128 von 1997/2000 als gleichermaßen gestaltbestimmendes wie die
Gestalt auflösendes Material einsetzt. Dieses Haus steht für sich und
zugleich unverkennbar in der Tradition der Moderne. Das können auch
andere, organisch geschwungene Glasformen wie das trotz seiner Glas-
fassade kryptische Neanderthal-Museum von Günter Zamp Kelp, Julius
Krauss und Arno Brandlhuber von 1996 oder das amöbenhafte Photonikzentrum in Berlin von Matthias Sauerbruch und Louisa Hutton aus
dem Jahre 1998. Es ist nicht die schwingende Grundform, sondern es ist
die Gliederung zweier getrennter, sich aber sehr nah kommender Bau-
körper, ihre räumliche Struktur, ihr Verhältnis zueinander und vor allem
ihre disziplinierte Fassade, die trotz Transparenz und von innen durchscheinender, farbig differenzierter Betonstützen wie selbstverständlich
als Einheit wirkt und den Körpern Volumen gibt.
Neue Oberflächen in der Architektur können offenbar nur dann lang-
fristig akzeptable Ergebnisse liefern, wenn sie nicht allein auf den Effekt
setzen, sondern ein Grundverständnis zwischen Funktion und Form wei-
terbestehen lassen. Soll hingegen die Oberfläche in erster Linie »Unver-
wechselbarkeit« erzeugen, »die das Bauvolumen nicht leisten kann«,
dann ist »mit solchen Identitätserwartungen (...) die Epidermis mehr
als überfordert, wie tätowierte Haut, die den Körper, den sie überspannt,
auch nicht vergessen lässt. Auch eine bedruckte Kiste ist eine Kiste.«[8]
In der gegenwärtigen Architektur kommt der Fassade eine immer größere Bedeutung zu, sie wird mehr noch als in der Phase der klassischen
Moderne zum »Manifest«.[9]
Oft ist dies aber nicht eine Reflexion veränderter Rahmenbedingun-
gen des Entwurfs und seiner theoretischen Grundlagen, sondern Teil
einer Vermarktungsstrategie. Denn es ist das Äußere eines Gebäudes,
das seinen Wahrnehmungswert und seine mediale Vermittelbarkeit
bestimmt.

Werner Sobek's House R 128 in Stuttgart (1997/2000). The latter is both
a complete one-off and yet is unmistakably in the Modernist tradition.
The same can be said of other organically curved glass shapes, such as
the Neanderthal Museum by Günter Zamp Kelp, Julius Krauss and Arno
Brandlhuber (1996), which despite its glass façade has a cryptic appearance, or the amoeba-like Photonics Centre in Berlin by Matthias Sauerbruch and Louisa Hutton (1998). It is not the curved plan form, but the
articulation of the two volumes that are separate but very close to each
other, their three-dimensional structure, their relationship, and above all
the disciplined façade that despite its transparency (which gives glimpses
of the different coloured concrete interior columns) makes the buildings
look quite naturally like a single entity and creates a sense of volume.
New surfaces in architecture are seemingly only able to produce results
that are acceptable in the long term if they are not solely concerned with
effect but also allow a fundamental understanding between function
and form to continue. If, by contrast, the surface is primarily intended to
create "a distinctiveness that the volume of the building is not able to,
that is placing an identity-creating expectation on it, which, as an epidermis, it cannot possibly fulfil; it is like tattooed skin that makes the
body it covers unforgettable. A box wearing a pretty print is still a box."
(Wolfgang Pehnt)[8] Façades are gaining ever more significance in contemporary architecture; to a greater extent even than in the period of Classic Modernism they are becoming a "manifesto."[9] However, this is often
not a reflection of the different conditions surrounding a design and its
theoretical basis but merely part of a marketing strategy. Because it is a
building's external appearance that determines how it is perceived and
how media-genic it is. The conditions under which architecture is created are too complex and complicated to be able to describe them simply
to the outside world. Symbols have multiple, often contradictory, mean-
ings that can be interpreted in different ways or sometimes not at all. The
loss of meaning can be beneficial to architecture, to its diversity, scope
for experiment and new ways of appropriating it, and the architecture
of surfaces fills the void that is left.
Thus, it is scarcely a surprise that the literature on this topic does not go
beyond technical treatments conveying basic technical knowledge that

Zamp Kelp, Krauss, Brandlhuber: Neanderthal–Museum, Mettmann, 1996

Matthias Sauerbruch, Louisa Hutton: Photonikzentrum Photonics Centre, Berlin, 1998

Die Entstehungsbedingungen von Architektur sind zu komplex und unübersichtlich, als dass sie ohne weiteres nach außen dargestellt werden können. Zeichen sind mehrfach, auch widersprüchlich belegt und lassen sich unterschiedlich oder auch überhaupt nicht deuten. Der Verlust von Bedeutungen kann auch ein Gewinn für die Architektur, für Vielfalt, Experimentierräume und neue Formen ihrer Aneignung sein, und in dieses Vakuum stoßen Oberflächen-Architekturen.

So verwundert es kaum, dass sich die Literatur zu diesem Thema in fachlichen Abhandlungen erschöpft, die das technische Basiswissen vermitteln, welches sich seit den siebziger Jahren nicht mehr grundlegend erweitert hat. Die Konstruktionen bei konventionellen Bauten sind bekannt und ausgereift.[10] Jüngere Titel, die scheinbar das gesamte Spektrum heutiger Möglichkeiten abbilden, entpuppen sich bei genauerer Betrachtung als die aufdringliche Leistungsschau eines Marktführers für Fassadensysteme, der ohne inhaltliche Bewertung qualitätvolle Architekturen gleichberechtigt neben Beispiele des Mittelmaßes stellt, weil offenbar beide mit seinen Produkten arbeiten.[11]

Es geht nicht mehr um die Technik, Fassaden zu konstruieren, sondern um das Material, aus dem sie bestehen. Interessant werden Architekturen, wenn sie den gedanklichen Sprung weg von tradierten Fügungen, Schichtungen, Verkleidungen in ihr theoretisches Konzept integrieren und die Betrachter und Nutzer an dieser Überlegung teilhaben lassen. Werden neue, neu kumulierte oder auch alte, wiedergefundene Materialien unreflektiert eingesetzt, dann können sie diese Aufgabe nicht übernehmen, dann sind die Ergebnisse meistens banal und immer beliebig. Wenn sie aber selbst zum Bedeutungsträger werden und ihre Aussage durch das bewusst eingesetzte »falsche« Material, die »sinnlose« Oberfläche überprüfbar machen, dann können sie der Architektur selbst starke neue Impulse geben. Die »Oberflächlichkeit« solcher Interventionen ist dann kein »Fassadismus«, sondern zeigt ein eigenes architektonisches Programm, weil sie den Charakter der Gebäude ausmachen, denn »Fassaden sind keine Stilfrage.«[12]

Innerhalb dieser Richtung sind für die Architektur besonders diejenigen Materialien interessant, deren serielle Herstellung sie verlässlich, dauerhaft und kostengünstig macht und die unabhängig von ihrer tatsächli-

has not fundamentally expanded since the 1970s. The structures used in conventional buildings are well known and fully developed.[10] More recent works, whose titles suggest that they cover the entire spectrum of current possibilities, turn out on closer examination to be nothing more than the flagrant display of achievements of a market leader in façade systems, who places high-quality architecture alongside a mediocre example without comment simply because both projects worked with his products.[11]

It is no longer the technology used to construct façades that is interesting but the materials they are made of. Architecture becomes interesting when it integrates into its theories the conceptual leap away from traditional compositions, layering, and claddings, and allows observers and users to share in these thoughts. If new, newly combined and cumulative materials – or even old, rediscovered materials – are used without reflection, they cannot fulfil this role; the results are usually banal and arbitrary. But if they themselves become the carriers of meaning and make that open to analysis by deliberating using the "wrong" materials or "meaningless" surfaces, then they can provide powerful new impetus for architecture itself. The fact that interventions of this kind relate only to the surface is then not just "facadism," but demonstrates that the surface has its own architectural programme because it has a decisive influence on the character of buildings. "Façades are not merely a question of style."[12]

Within this approach, the materials that are particularly interesting for architecture are those that can be mass-produced, making them reliable, durable and good value for money. Also highly valued are those that, irrespective of what they are actually made, have core or top layers that are coloured, printed, machined or sealed in such a way as to depict a particular material, imitate it or even look more natural than the material itself. Particularly synthetic processes make it possible to highlight certain desirable properties that make the product look more real than the real thing. It is an intentional effect, which is found not only as a stylistic characteristic in small-scale contexts in architecture's "off-scene," but which is increasingly characterising both experimental projects and projects that conform with market expectations that are the work of established architects.

Hild und K Architekten: Haus mit Sozialwohnungen Social housing block, Kempten, 1997

chen materiellen Substanz mit ihren Träger- und Deckschichten bis hin zu farbigen, bedruckten, gefrästen, versiegelten Oberflächen einen vorher bestimmten Stoff darstellen, ihn imitieren oder in seiner natürlichen Wirkung sogar übertreffen. Vor allem synthetische Verfahren erlauben es, bestimmte gewünschte Eigenschaften hervorzuheben, die das Produkt wahrer als die Wirklichkeit erscheinen lassen. Es ist ein gewollter Effekt, der nicht nur als Stilmerkmal bei Kleinformen der Off-Architekturszene zu finden ist, sondern der zunehmend auch gleichermaßen experimentelle wie marktkonforme Projekte etablierter Architekten prägt.

Die archaische Wand, die schützt, trägt und deren Material für sich spricht, gibt es nicht mehr. Sie ist längst ein leichtes, variables und funktionalisiertes System wie die Moderne selbst. Nur ihre äußere Schicht ist ästhetisch wirksam. Es gibt kaum noch einen Bereich des Bauens, in dem nicht farbige, glatte oder strukturierte Platten aus den unterschiedlichsten Materialien mit den unterschiedlichsten Oberflächenanmutungen elementiert als Fassadenverkleidung verwendet werden.

Architekten wie Hild und K, die eine besondere Affinität zum provozierenden Experiment mit Fassaden haben, verwendeten 1997 dunkle Phenolharzplatten mit quer verlaufendem Mahagonidekor, um die Fassade eines Hauses mit 12 Sozialwohnungen in Kempten zu verkleiden – gegliedert und rhythmisiert durch Aluminiumformteile und die sichtbaren Verschraubungen der Platten. Da es unmöglich ist, Platten dieser Größe und Struktur aus echtem Holz herzustellen, geht es um ein Spiel mit der Wahrnehmung, die Brechung von Sehgewohnheiten und den produktiven Kontrast zu älteren Nachbarhäusern, deren Außenwände vollständig mit grauen, kleinteiligen Fassadenplatten vermutlich aus Faserzement verkleidet sind.

Resopal, einer der bis in den Sprachgebrauch hinein Alltagskultur stiftenden Kunststoffe für Interieurs der fünfziger Jahre und später lange Zeit Kurzformel und Synonym für Besonderheiten jener Zeit mit ihrem Hygienetick, ihrer vermuteten Oberflächlichkeit, ihrem Hunger nach Farben und modernen Materialien, hat längst einen Imagewandel vollzogen. Es gilt inzwischen als hochwertig und teuer und findet als Resoplan zunehmend auch den Weg an die Fassaden von Gebäuden. Und dies in allen denkbaren Dekoren bis hin zur scheinbaren Auflösung der Materie durch

The archaic wall, which protects, carries loads and is made of materials that speak for themselves, no longer exists. For a long time the wall has been a light, variable and functionalised system – like Modernism itself. Only its outer layer has an aesthetic effect. There is scarcely a single area in architecture in which coloured, smooth or textured panels, made of the most widely differing materials with surfaces seeking to represent an equally broad range of materials, are not being used as elements of façade cladding.

In 1997, architects such as Hild and K, who have a particular affinity for provocative experiments with façades, used dark phenolic resin panels with horizontal mahogany patterning to clad the façade of a 12-unit social housing block in Kempten, with articulation and rhythm provided by moulded aluminium components and the visible bolts fixing the panels. Since it is not possible to make panels of this size and structure from real wood this is a play with our perception, breaking habitual ways of seeing things and providing a productive contrast with older neighbouring buildings, their external walls completely clad in small, grey façade panels that are probably made of fibre cement.

Resopal – one of the synthetic materials for interiors developed in the 1950s that shaped everyday life and went into the language itself, later becoming a synonym for the peculiarities of the time and its obsession with hygiene, its alleged superficiality, its hunger for colour and modern materials – has long since undergone an image makeover, and it is now seen as an expensive up-market material. In the form of Resoplan, it is increasingly finding its way onto the façades of buildings in every conceivable pattern including the seeming dissolution of material through surfaces that appear to be out of focus when the light falls on them at a particular angle. They are intended to do the exact opposite of what architecture usually does: they abolish the defining edges of spaces. These interventions move fundamentally in a milieu that has Modernist leanings, and they can only work if they encounter aesthetic education, prior cultural understanding and an eagerness to experiment on the part of the client.

The material of surfaces becomes interesting precisely at the point where it abandons its historical connotations and appears as a virtual building

unscharf wirkende, unter einem bestimmten Einfallswinkel des Lichts beinahe schon immaterialisierte Oberflächen, die genau das Gegenteil von dem erreichen sollen, das Architektur normalerweise leistet: Sie entgrenzen den Raum. Diese Interventionen bewegen sich grundsätzlich in einem modernistischen Umfeld, und sie können nur funktionieren, wenn sie auf ästhetische Bildung, auf kulturelles Vorverständnis und auf Experimentierfreudigkeit auch seitens der Auftraggeber treffen.

Das Material der Oberfläche wird genau dort interessant, wo es in doppelter Brechung seine architekturhistorischen Konnotationen verlässt und als virtueller, gleichermaßen individualisierter wie standardisierter Baustoff in Erscheinung tritt. Er verhüllt die Form in einem technischen Sinne und legt die Bedeutung dieses Tuns zugleich offen, weil die Materialimitation natürlich sofort zu erkennen ist. Es ist eine neue Art der Trompe-l´Œil-Architektur, die dem Material ohne Eigenschaften eine Möglichkeitsform gibt. »Zeitlos« sind diese Architekturen, die nicht altern können, nicht. Auch wenn das Material selbst über der Zeit steht, ist seine Auswahl, seine Fügung und seine Rolle zeitgebunden. Und dies wird es erlauben, das Haus später einmal jahrzehntgenau zuzuordnen.

material that is individualized and standardised to equal degrees. It covers up the building's form in a technical sense, at the same time exposing the meaning of this action, because it is, of course, instantly recognisable as an imitation. It is a new form of *trompe-l´œil* architecture that gives possible form to a material that is essentially featureless. This kind of architecture that cannot age is nevertheless not "timeless." Even if the material itself endures over time, its selection, arrangement and role are linked to a particular time and will make it possible in years to come to date the building to the precise decade.

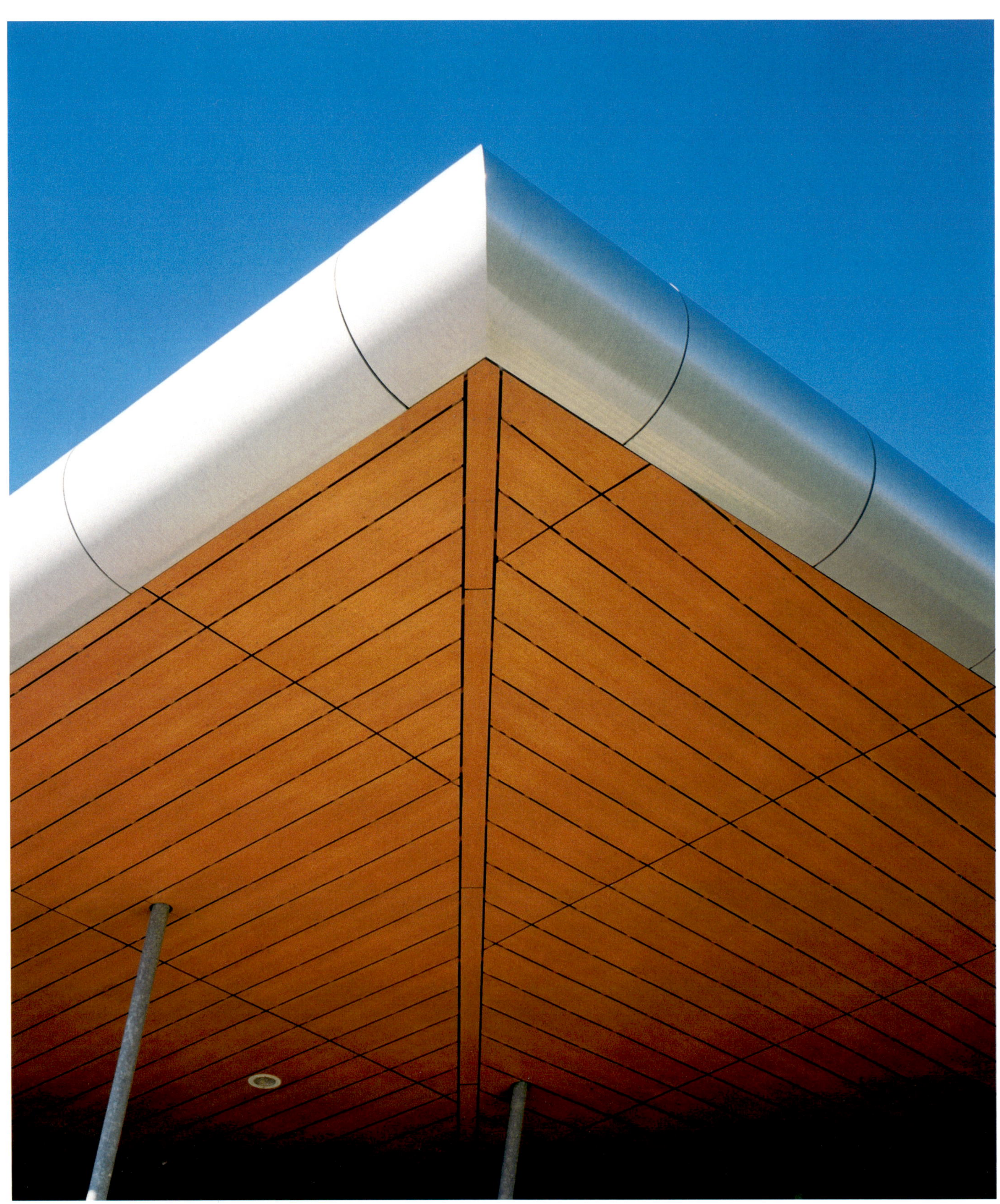

Dach aus Resoplanplatten, REIGERBOYS-Stadion, Heerhugowaard, 2003
Roof in Resoplan panels, REIGERBOYS Stadium, Heerhugowaard, 2003

RESOPAL UND ARCHITEKTUR
RESOPAL AND ARCHITECTURE

Der Einsatz der Resopalplatte in Außenbereichen unter dem Namen Resoplan begann Mitte der siebziger Jahre. Vorausgegangen waren eine jahrelange Entwicklungsphase und die praxisnahe Erprobung auch unter extremen Klimabedingungen, durchgeführt vom Hersteller Resopal Werk H. Römmler GmbH und von unabhängigen Instituten. Größere Bauprojekte wurden von den Behörden in den ersten Jahren allerdings nur in Einzelfällen genehmigt. Wie bei allen neuen, noch nicht genormten Baustoffen bedurfte es erst einer Zulassung der Bauaufsicht. Nach zahlreichen Überprüfungen durch die Behörden und vielen eingereichten Nachweisen erteilte 1983 das Berliner Institut für Bautechnik die Erlaubnis für den baulichen Einsatz.

Mit der Genehmigung fand die Resoplanplatte schnell eine weite Verbreitung vor allem bei der Verkleidung von Fassaden und Balkonen. Insbesondere für Zweckbauten und im Wohnungsbau war der Einsatz dieses Werkstoffs höchst willkommen, denn die ihm eigene Witterungsbeständigkeit, der geringe Pflegeaufwand und der wegfallende Neuanstrich verringerten die Instandhaltungskosten beträchtlich.

Diese reinen Nutzgedanken für die Anwendung von Kunststoffplatten, die bei den meisten Gebäuden zu Sterilität und Anonymität führen, machen den Einsatz des Materials für viele Architekten, denen es auf Baukunst ankommt, immer noch schwer. Allerdings scheint sich in den letzten Jahren auch unter diesen, eine unbefangenere Haltung bei der Fassadengestaltung durchzusetzen, die das kreative Spiel mit Modischem, Konventionellem und Historischem ausdrücklich einschließt. Die neueren Beispiele, die im Folgenden vorgestellt werden, fügen sich allesamt in einen baulichen Kontext, der durch diese Kriterien bestimmt wird. Die drei Gebäude – Mietshaus, Schule und Wohnheim – sind durch ihre Funktion geradezu prädestiniert für die Verwendung von pflegeleichten Kunststoffplatten. Allerdings werden sie hier nicht als simple Verkleidung, sondern als architektonisches Gestaltungsmittel eingesetzt. Bei dem Wohnhaus konnte mit »Resopal pur« sogar dem Wunsch nach einer »natürlichen Erscheinung und Oberfläche« entsprochen werden. Wie im Fall der Schule nutzten die Architekten auch hier die Möglichkeit, dieselbe Oberfläche – in unterschiedlicher Qualität der Resopalplatte – innen und außen einzusetzen.

The use of Resoplan, the Resopal laminate designed for exteriors, began in the mid-seventies. This was preceded by years of development and practical trials, including testing under extreme climate conditions conducted by the manufacturer Resopal Werk H. Römmler GmbH and by independent institutes. However, in the early years its use in larger building projects was approved by the authorities only in isolated cases. As with all new building materials before they are standardised, the building control agency first had to approve the product. After numerous tests by government authorities and submission of a host of documentary evidence, the Berlin Institute of Building Technology approved the use of the product for construction purposes in 1983.

This accelerated the application of Resoplan and it soon became widespread, particularly in cladding for façades and balconies. The use of this material was extremely welcome in functional buildings and housing, since the fact that it was weather-resistant, low-maintenance, and did not need repainting reduced costs significantly.

It is this purely utilitarian rationale behind the use of plastic panels that gives most buildings a look of sterility and anonymity making it difficult for many architects who are interested in creating high-calibre architecture to use the material. However, in recent years such architects also seem to have adopted a more open-minded approach to façade design, which includes creative play with trendy, conventional and historical elements. The more recent examples of the use of Resoplan illustrated below all fit into an architectural context that is characterised by these criteria. The function of the three buildings – a block of flats, a school and a care home for the elderly – makes them predestined to use these easy-maintenance plastic panels. However here, they are not just a functional cladding but are also an architectural design feature. In the case of the block of flats, it was even possible to use "pure Resopal" to fulfil the client's wish for a "natural appearance and surface finish." As in the case of the school, the architects here also took the opportunity to use the same surface finish on the inside and outside of the building, using different grades of Resopal.

■ **FASSADENVERKLEIDUNGEN MIT RESOPLAN**
Beispiele aus den siebziger und achtziger Jahren

■ RESOPLAN FAÇADE CLADDING
Examples from the seventies and eighties

■ **HILD UND K ARCHITEKTEN: MIETSHAUS, KEMPTEN, 1997**

Das Gebäude mit zwölf Sozialwohnungen sollte auf Wunsch des Bauherrn eine Verkleidung aus Kunststoffplatten erhalten. Um dessen Materialität zu betonen, wurde ein Dekor gewählt, das seinerseits auf ein Material verweist. Durch die »Unmöglichkeit« von Mahagoniplatten dieser Größe wird die künstliche Materialität der Fassade überhöht. Damit fügt sie sich in die mit den unterschiedlichsten Verkleidungen versehenen Häuser der Umgebung ein.

■ **HILD UND K ARCHITEKTEN: BLOCK OF FLATS, KEMPTEN, 1997**

The social housing complex containing twelve flats was clad with plastic panels at the client's request. To emphasize the particular quality of the synthetic material, a decorative paper was selected that itself suggests a material. The "impossibility" of mahogany sheets of this size points up the synthetic quality of the façade, fitting in with the houses in the area and their great variety of cladding.

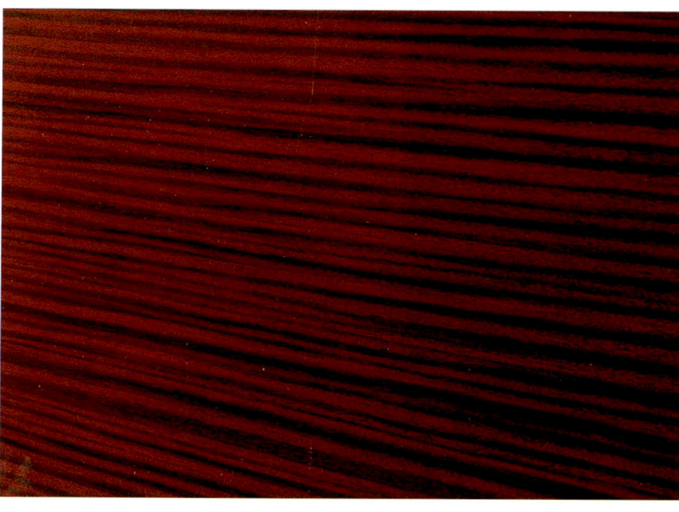

Um die Oberflächengestaltung von außen nach innen fortzuführen, wurde das Material Resopal ausgewählt, weil verschiedene Produkte bei gleicher Farbe und sehr ähnlicher Oberfläche unterschiedliche Funktionen erfüllen konnten. Die Innenverkleidung musste zum Beispiel auch schallabsorbierend sein. Die Platten wurden in horizontalen Streifen mit gleicher Höhe, aber verschiedenen Längen verlegt. Dabei wurde jeder Farbe ein Format zugewiesen.

■ NIEMANN ARCHITEKTEN: EXTENSION TO THE SCHWARZENBERG-STRASSE SPECIAL SCHOOL, HAMBURG, 2005

In order to continue the design of the exterior of the building through to the interiors, Resopal was chosen, because different products performing different functions can have the same colour and very similar finishes. For example, the interior panelling had to also be sound absorbent. The panels were arranged in horizontal strips that were identical in height but different in length. Each different colour had a different format.

■ **CLAUS EN KAAN ARCHITECTEN: LEO POLAKHUIS,
AMSTERDAM, 2006**

Das Zentrum zur Betreuung älterer Menschen wird von einem sechzehngeschossigen Wohnhochhaus dominiert. Insgesamt drei Gebäude mit jeweils drei Geschossen sind ihm vorgelagert. In ihnen befinden sich medizinische und therapeutische Einrichtungen, eine Tagesstätte sowie ein Restaurant. Nur diese Bauten, die allen Bewohnern dienen sollen, sind mit Resoplan verkleidet. Die heiteren Farben und begrünten Höfe schaffen eine angenehme Atmosphäre.

■ **CLAUS EN KAAN ARCHITECTEN: LEO POLAK HOUSE,
AMSTERDAM, 2006**

This elderly care complex is dominated by a 16-storey block of flats. In front of it are three three-storey buildings, housing medical and therapeutic facilities, a daycare centre and a restaurant. Only these buildings, which are intended for use by all the residents, are clad in Resoplan. The bright colours and planted courtyards create a pleasant atmosphere.

Ziel war, beim Entwurf des zweigeschossigen Gebäudes zwei auf dem Nachbarareal befindliche Trockenscheunen zu berücksichtigen. Bei der Wahl des Materials kam nur ein »echter« Werkstoff in Frage, der eine natürliche Erscheinung aufweist und für den Einsatz innen und außen geeignet ist. Durch die Verwendung der braunen Resoplanplatte konnte die Klarheit und das Fugenbild des Baukörpers herausgestellt und der gewünschte Kontrast zwischen Alt und Neu erreicht werden.

The design aim of this two-storey building was to take into account two drying barns on neighbouring land. In the choice of materials, there was no question of anything other than an "authentic" material that looked natural and was suitable for both indoor and outdoor use. By using the brown Resoplan panel, it was possible to emphasise the building's clarity and joint pattern and achieve the desired contrast between old and new.

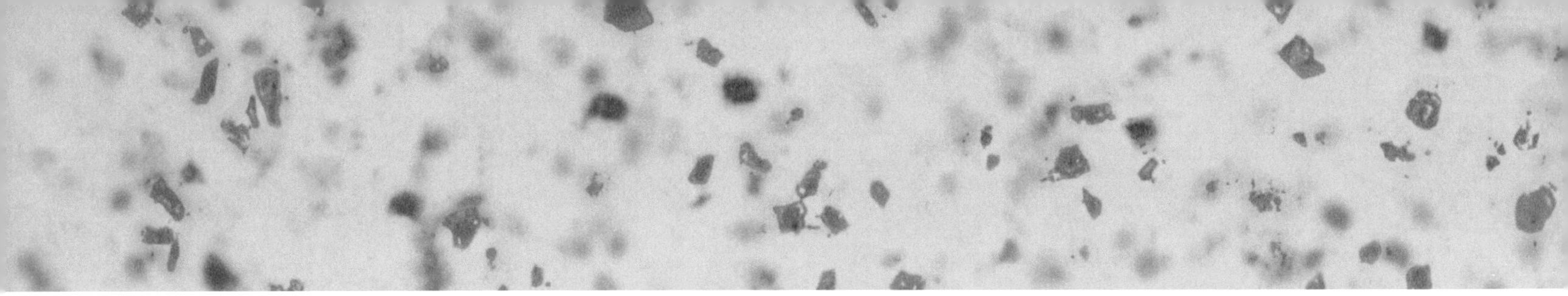

Gert Selle
RESOPAL – DAS OBERFLÄCHENPHÄNOMEN
RESOPAL – THE SURFACE PHENOMENON

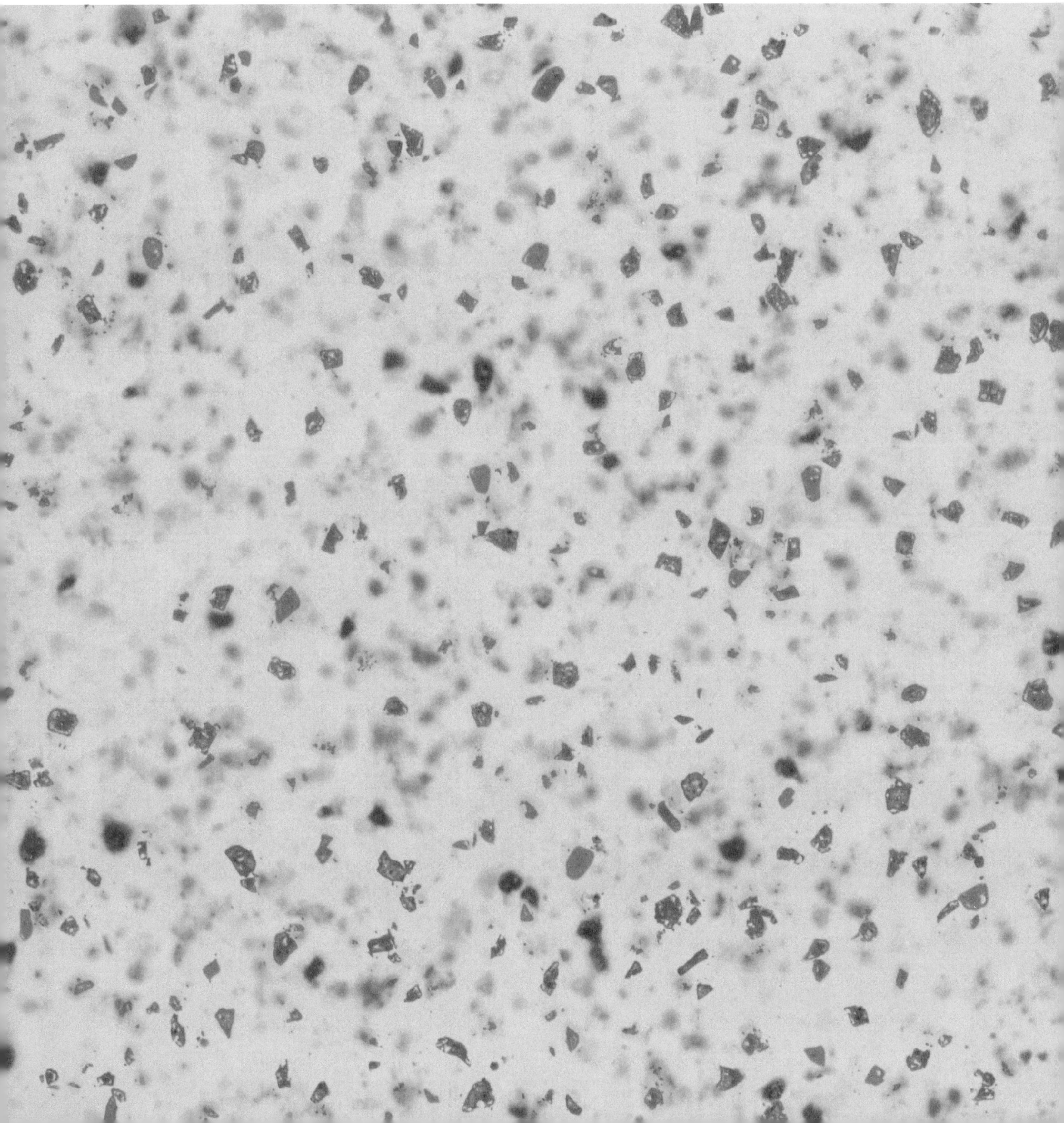

In den Alltag integriert, begleitet dieses Produkt die jüngere Kulturge-schichte des Industriezeitalters als ein stiller Indikator fortschreitender Immaterialisierung. Undurchdringlich bedeckt es Flächen wie eine iso-lierende Schicht, ja möchte als Oberfläche an sich verstanden sein.

Im Anwendungsfall verschwindet sein millimeterdünner Körper ganz und wird zur Fläche. Resopal ist ein Zwitter in mehrfacher Hinsicht. Man weiß nicht, ob man es der Architektur- oder der Designgeschichte zuord-nen soll. Darüber entscheidet erst die spezielle Nutzung. Resopal ist ein duroplastisches Halbzeug, das zur Wirkung kommt, sobald es anderen Flächen aufgebracht wird. Zur Bekleidung von Möbeloberflächen, Wän-den und Hausfassaden gedacht, soll es deren Oberflächen Eigenschaf-ten verleihen wie Härte, Makellosigkeit, Unempfindlichkeit, Homogeni-tät und Glätte.

Das Plattenprodukt entsteht aus phenolharz-imprägnierten Kernpapie-ren, unter Hitze und starkem Druck mit einem Dekor-Papier und einem durchsichtigen Overlay verbunden. Es handelt sich also um eine Verbin-dung von Kunststoff und Naturmaterial. Die Herstellung bewegt sich zwischen Manufaktur (»Bogenlegerinnen« stapeln per Hand die Bögen und legen als letzte Schicht das Dekorpapier auf) und automatisierten Fertigungsstrecken.

Wie die resopalbeschichtete Pressspan- oder Tischlerplatte eine undurch-lässige Oberfläche hat und diese im Anwendungsfall sichtbar nach außen gekehrt zeigt, hat auch die dünne Resopalschicht ohne Trägeruntergrund eine Schauseite und eine Kehrseite zum Aufkleben auf ein anderes Mate-rial. Die Schauseite bleibt entweder neutral oder bietet sich für dekorati-ve Gestaltungen an. Nach seiner immanenten Logik ist Resopal ein End-losprodukt, das sich in seiner Zweckbestimmung über Flächen ausdehnt wie Asphalt oder Beton. Im Gegensatz zu deren massiv körperlicher Prä-senz tritt Resopal als reine Fläche, scheinbar ohne Volumen und Masse auf. Seine generelle Aufgabe als Oberflächenbedeckung ist es, eine Gren-ze zwischen dem Kompakt-Festen und dem Raum zu markieren. In die-ser Funktion hat sich die Resopalplatte bis heute bewährt. Sie begrenzt und schließt ab, selbst dort, wo sie sich in neuen Inszenierungen schein-bar zum Raum öffnet.

Man hatte sich an die neutralen Flächenbeschichtungen daheim, am Arbeitsplatz oder in öffentlichen Räumen und Verkehrsmitteln gewöhnt. Die Versiegelung entzog sich der Aufmerksamkeit, weil man mit ihr immer irgendwie konfrontiert war. Bemerkt wurde der glatte Press-stoffbelag eigentlich nur bei Einführung der Küchenhängeschränke mit ihren pastellfarbenen Schiebetüren oder auf den legendären Nie-ren- bzw. Palettentischen der fünfziger Jahre. Sein Eroberungszug über die Fläche folgte in aller Stille. Heute tritt die Resopalplatte aus dem Schatten ihrer Unauffälligkeit heraus und macht neue Wahrnehmungs-angebote.

Geschichtlich laufen zunächst zwei Produkttypen parallel: betont kör-perhaft-plastisch ausgeformte Gebrauchsgegenstände aus Bakelit bzw. dem Nachfolgewerkstoff Aminoplast und die Hartpapier-Pressstoffplat-te als flächig formatiertes Halbzeug, das später, unter der Markenbe-zeichnung Resopal (in der DDR Sprelacart) perfektioniert und diversi-fiziert, eine kulturell flächendeckende Bedeutung erlangt. Wer heute Resopal sagt, meint die Platte und vergisst dabei, dass es verwechselba-re Billigprodukte gibt, bei denen nur ein einziger Papierbogen auf eine Spanplatte aufgepresst ist.

Dass es anfangs zwei Arten der Werkstoffverarbeitung gegeben hat, ja sogar noch einmal, als das Unternehmen nach dem Krieg im Wes-ten wieder produzierte, hat man schon vergessen. Die Linie der skulptu-ralen Objekte wurde vom zweidimensionalen Flächenbedeckungspro-

Integrated into everyday life, this product accompanies the recent cultural history of the industrial age as a silent indicator of progressive demateri-alisation. Impenetrable, it covers surfaces like an insulating layer; indeed it would like to be understood as a surface in its own right.

In use, its slender profile vanishes completely and becomes two-dimen-sional. Resopal is a hybrid form in several respects. It is impossible to know whether it should be classified under history of architecture or history of design; it is the specific use that answers that question. Resopal is a semi-finished duroplastic product that becomes effective as soon as it is applied to other surfaces. Intended as a surface finish for furniture, walls and building facades, the idea is that it should give the object to be cov-ered, or rather its surface, particular properties such as hardness, flaw-lessness, resistance, homogeneity and smoothness.

The laminated product is made up of core layers of paper impregnated with phenolic resin, topped by printed decorative paper with a transpar-ent melamine resin-impregnated overlay fused together under heat and high pressure. Thus, the final product is a combination of natural and arti-ficial materials. The manufacturing process is also a combination of auto-mated production lines and manual skills (the core and decorative sheets of paper are layered manually).

Just as the particle board or block board onto which Resopal is laminat-ed have an impermeable surface which is the side that is visible when the product is in use, the thin sheet of Resopal before it is applied to a substrate also has a display side and a reverse side, which is bonded to another material. The display side either remains neutral or can be decorated. The inherent logic of Resopal is that it is an "endless" prod-uct, whose purpose is to cover surfaces just as asphalt or concrete does. However, by contrast with the solid physical presence of those materi-als, Resopal is purely two-dimensional, apparently without volume or mass. Its general function as a surface finish is to demarcate a bounda-ry between what is compact and solid, and space. The Resopal board ful-fils this function well. It gives a sense of closure and demarcation, even in applications where it has been used in new orchestrations that seem-ingly open it up to a space.

We had become accustomed to neutral surface finishes at home, at work, or in public spaces and means of transport. The idea of a sealant no long-er attracted our attention because we were somehow constantly con-fronted with it. The smooth laminate covering was never actually noticed until kitchen wall cabinets with their pastel-coloured sliding doors were introduced, or when it made its appearance on the legendary *Nieren-tisch,* the kidney-shaped table so popular in the 1950s. It then stealth-ily continued its conquest of the surface. Today, the Resopal laminate is emerging from its own shadow of inconspicuousness and offering us new takes on perception.

Historically speaking, two types of product run in parallel: emphatical-ly physical, fully shaped everyday articles made of Bakelite or its succes-sor material aminoplast, and the laminated board as a two-dimensional semi-finished product, which was later perfected and diversified under the brand name Resopal (Sprelacart in East Germany) and acquired a broad-based cultural significance. Anyone who talks of Resopal today means the board, forgetting that cheap look-alikes exist, which are noth-ing but a single sheet of paper fused onto chipboard.

It has been forgotten that there were initially two phenotypes of process-ing this material, and that they were even produced again after the War when the company set up in the West. The sculptural object line was replaced by the two-dimensional surface finishing product, because three-dimensional objects could now be produced more easily and

Frühestes Beispiel für Postforming, Wilson House, 1959
Earliest example of postforming, Wilson House, 1959

dukt abgelöst, weil sich anderenorts plastische Objektformen aus neuen Kunststoffen leichter und billiger herstellen ließen und es in der Möbelindustrie und beim Innenausbau zunehmend Bedarf für das vielseitig verwendbare Plattenmaterial gab. Freilich entspricht die Entscheidung des Unternehmens, nur das Plattenprodukt weiterzuentwickeln und die Herstellung dreidimensionaler Objekte zu beenden, einem historischen Trend, den man als langfristigen Prozess der Auflösung des Produktkörpers zugunsten von Benutzeroberflächen beschreiben kann.

Mit der Resopalplatte assoziiert man eine über Jahrzehnte anhaltende funktional-nüchterne Ausstattungskultur, verbunden mit einer aseptischen Ästhetik des Raumes. Hinter dem Inbegriff des Praktischen, Sauberen, Dauerhaften und Neutralen verbirgt sich eine unbewusst mitvollzogene reduktionistische Umorientierung sinnlicher Erfahrung. Wurde bei den Bakelit-Objekten der frühen dreißiger Jahre Phenoplastpulver unter Druck und Hitze zu einer festen Masse verdichtet, wobei zuvor das Formwerkzeug nach einem vollplastischen Modell hergestellt werden musste, wird die Kunstharz-Pressstoffplatte in ihrer Flächenschichtung lediglich unter Druck und Hitze zwischen planen Stahlblechen erzeugt, anderes Formwerkzeug entfällt – es geht ja nicht mehr um Körper.

Im ersten Fall waren Designer als plastizierende Künstler gefragt wie Christian Dell, der farbiges Picknickgeschirr aus dem 1930 zum Patent angemeldeten Resopal entwarf. Um 1931 soll die Bordküche für einen der Zeppeline aus dem neuen Werkstoff produziert worden sein – ein Beispiel wahrscheinlich nur flächiger Anwendung des Materials. Bei den Objekten wird die plastische Präsenz im Raum betont. Die Platte hingegen verliert ihren Körper in der Anwendung. Die Resopal-Massivplatte, die maximal bis zu 42 mm stark ist und als freitragende Tischplatte dienen kann, scheint eine Ausnahme. Sie ist aber letztlich auch nur eine Fläche, die ihren Träger gleich mitbringt. Das normale Plattenprodukt verfügt über keine Masse, die plastisch modelliert werden kann. Das Material lässt sich beim sogenannten Postforming nur an den Kanten unter Wärme biegen, solange es noch keinem Träger aufgebracht ist. Eine Ummantelung sphärisch gekrümmter Körper ist nicht möglich. Die Platte zwingt demnach die Verarbeitungsphantasie in die Ebene: Glatte Wände oder die Seiten eines Kubus bieten sich an. Grundcharakteristikum des Materials

cheaply elsewhere using new kinds of plastic and because there was a rising demand for this versatile board product in the furniture industry and interior fittings. Admittedly the company's decision to develop only the board and cease production of the three-dimensional objects was in line with a historical trend that can be described as a long-term process by which the three-dimensional quality of products gradually disappeared and was replaced by the "user interface."

With the Resopal laminate we associate a functional, no-frills furnishing culture that lasted for decades, linked to an aseptic spatial aesthetic. Behind the idea of practicality, cleanliness, durability and neutrality lies an unconscious reductionist reorientation of sensory experience. The Bakelite objects of the early 1930s were made by compressing powdered phenolic resin under heat and pressure to form a solid mass, the mould having been made prior to this from a fully shaped model. The laminated board was produced simply under heat and pressure between two flat sheets of steel; no other mould was required – it was no longer a question of producing three-dimensional objects.

In the first case, designers were in demand as plastic artists. We recall the coloured Resopal picnic set designed by Christian Dell, for which a patent application had been filed in 1930; the galley for one of the Zeppelins is thought to have been created in the new material around 1931, an example of the material being used two-dimensionally. In the case of objects, the three-dimensional presence is emphasised. By contrast, the laminated board loses all appearance of physicality when it is used. The solid Resopal board, which can have a thickness of up to 42 mm and can also be used, for example, as a cantilevered tabletop, seems to be an exception; but it, too, is only a surface that brings its substrate along with it. The normal board product does not have any mass that could be plastically shaped. It is only possible to bend the material by so-called post-forming before it has been applied to a substrate. It cannot cover spherical objects. The board thus forces the imagination of the designer into the two-dimensional; smooth walls or the sides of a cube are obvious choices for planes to be covered. The basic characteristic of the materials is its two-dimensional use; it determines the image perceived. Our visual and tactile senses tell us that Resopal is a flat surface, not a three-dimensional object.

Resopaloberfläche Blurington Silver, 2006
Resopal surface Blurington Silver, 2006

ist sein flächiger Einsatz, der das Wahrnehmungsbild prägt. Wir sehen und tasten Resopal als plane Oberfläche, nicht als Körper.

Davon konnte bei den von der H. Römmler AG in Spremberg gefertigten Telefon- und Radiogehäusen aus dem firmeneigenen Bakelit noch keine Rede sein. Hier durften sich Designer als plastische Formgeber profilieren und den Gerätebenutzern handgreiflich-körperliche Erfahrungen vermitteln. Heute finden sie sich als Dekor-Entwerfer an Computern wieder, um Oberflächengestaltung zu betreiben, weil das Produkt ihnen anderes nicht mehr abverlangt. Oder es gibt sie gar nicht mehr, weil die Dekorpapier-Druckereien selbst natürliche Materialien digital abtasten, um die Daten ihren Maschinen einzugeben. Im Prinzip kann heute jeder Auftraggeber einen Entwurf vorlegen, sofern er eine Mindestmenge Dekorpapier bestellt. Designer, Kunden und Drucker befassen sich gegenwärtig mit Motiven, die in ihrer simulierten Materialität zwischen Wirklichkeit und Illusion oszillieren. Die haptische Erfahrung tritt in den Hintergrund, wird vorrangig durch visuelle Reize ersetzt oder minimal durch gelegentlich der Oberfläche aufgepresste synthetische Materialstrukturen animiert. Alles Körperlich-Greifbare verlagert sich als Fake auf die Oberfläche, die zum Bild von etwas wird, das gar nicht da ist. So geht es in den neuen Resopal-Kollektionen um virtuelle Überbietung des Realen durch visuellen Hyperrealismus, während die einst plastisch ausgeformten Haushaltsgeräte und elektrotechnischen Fabrikate als Zeugnisse einer vergangenen skulpturalen Produktkultur zu begehrten Sammlerstücken aufgestiegen sind.

Heute ist die Produktion von Simulakren auf der Fläche angesagt. Man hätte nicht erwartet, dass die im Alltag lange kaum mehr beachtete Resopalplatte noch einmal Aufmerksamkeit erregen würde. Hinter ihr liegt eine Durststrecke der Reizarmut. Vor allem in Gestalt der weiß beschichteten Pressspanplatte hat sie sich selbst ein Denkmal der Eintönigkeit gesetzt. Sie hat ihren Rückhalt folgerichtig in der Ideologie des Funktionalismus gefunden, aber es wundert auch nicht, dass sie Aufwertungsversuchen ihrer Wahrnehmbarkeit nicht widerstanden hat. Ihrer materialästhetischen Sterilität war nur durch Farbgebung und Oberflächendekor beizukommen. Inzwischen kann das Unternehmen seinem Produkt exorbitante Wahrnehmungsqualitäten bescheinigen.

This could not have been said of the telephone and radio housings made by H. Römmler AG in Spremberg from the company's very own Bakelite. Here designers had the opportunity to upgrade their image as artists working in three dimensions and create visible and tangible experiences for users of the appliances. Today they find themselves sitting at their computers, back in the role of decorative artists engaged in designing surface finishes because that is all the product requires of them. Or they do not even exist any more, because the printers of decor paper scan and digitalise natural materials themselves so that they can feed the data into their printing machines. In principle, any client can now submit their own design, provided they order a minimum quantity of decor paper. Designers, customers and printers are currently concerned with motifs that vacillate in their simulated materiality between reality and illusion. Tactile experience is pushed into the background, but it is activated by visual stimuli or occasionally animated by synthetic material structures pressed onto the surface. Everything that is physical and tangible is transposed as a fake onto the surface where it becomes the image of something that is not actually there. For example, the new Resopal collections are all about virtually out-trumping the authentic by visual hyperrealism, while the once fully shaped household appliances and electrotechnical goods that testify to a bygone culture of sculptural products have become highly sought-after collector's pieces.

Today, it is all about producing surfaces that simulate reality. No-one expected that the Resopal laminate, which for a long time had been practically ignored in everyday life, would once more attract attention. It went through a long arid stretch when it was pretty much devoid of appeal. It has only itself to thank for erecting a monument to monotony in the shape of white laminated chipboard. It logically turned to functionalism for ideological support; but it is no wonder that it was not able to withstand the attempts to enhance its perceptual appeal. Its aesthetic sterility as a material could only be countered by colour and surface decoration. The company can now certify that its product has extraordinary perceptual qualities.

That was in doubt for a long time. An empirical study of the quality of experience provided by Resopal furniture, carried out in 1977 by psychol-

Kantine im Resopalwerk, siebziger Jahre
Canteen at the Resopal factory, 1970s

Lange bestanden daran Zweifel. Auf der Grundlage einer empirischen Studie zur Erlebnisqualität von Resopalmöbeln, die der Psychologe Friedrich Heubach 1977 durchgeführt hat, ergab sich zunächst ein schlüssiges Negativprofil des Materials. Heubach konstatierte eine »psychologisch defizitäre Gegenständlichkeit der Resopalmöbel«, die darin bestehe, dass der implizite Entwurf von Sinnlichkeit, der aus dem Umgang mit dem Material resultiere, eine »Enthistorisierung der sinnlichen Erfahrung« bewirke, weil Resopal »zeitlos« sei und die Spuren seiner Benutzung nicht bewahre. Darüber hinaus finde eine »Depersonalisierung« der Erfahrung statt, weil Nutzer an diesem Material keinen Anhalt zur »individuellen Besonderung« fänden. Schließlich sei eine »Partialisierung der sinnlichen Erfahrung« festzustellen, weil das Material sich ästhetischen Erwartungen versage, die über seine reine Funktionalität hinausgehen. Es fehle ihm das »anschauliche ›Mehr‹ des Ästhetischen, ein Moment von ›Stimmung‹‹.[1]
Damals waren diese Befunde valide und zustimmungsfähig. Unter dem Aspekt einer weitergehenden Kulturgeschichte der Materialien muss man jedoch den verblüffenden Schluss ziehen, dass die damaligen Beobachtungen inzwischen positiv auszulegen sind. Resopal, dieses Material ohne ausgeprägt sinnliche Eigenschaften, befindet sich heute auf der Höhe der Zeit, *weil* es sich durch Künstlichkeit, Alterslosigkeit und Neutralität der differenzierten sinnlichen Erfahrung und Qualifizierung entzieht. Es hat heute sehr wohl ein »anschauliches Mehr des Ästhetischen« zu bieten – in der scheinhaften Reproduktion dessen, was an ihm einst vermisst wurde.
Es ist das nüchtern-praktische, dauerhafte, gleichmäßige, synthetische, harte Oberflächenmaterial geblieben, darin einst erwiesenermaßen sachlich-modern und illusionslos funktionalistisch, was die kritischen Sensualisten auf den Plan rief. Inzwischen verhält es sich so konsequent postmodern, dass es sich umstandslos in ein Design der Trans-Sensualität, der Produktion virtueller Erfahrung einklinkt, aktuelle Bedürfnisse anspricht und in Gestalt perfektionierter Materialimitate geradezu zeitgeistverdächtig geworden ist. Die Mängel von einst sind zu Tugenden mutiert. Grundthema ist heute ja gerade das nicht mehr real Greifbare, sondern die Enthistorisierung aller Dinge und Oberflächen, der Daueraufenthalt in künstlichen Welten. Die Resopalplatte profitiert von die-

ogist Friedrich Heubach, found that the material had a decidedly negative image. Heubach observed that "Resopal furniture had psychological shortcomings as an object," which consisted in the fact that the innate sensory quality in the design, which results from the treatment of the material, "actually takes the sensory experience out of any historical context," because Resopal is "time-less" and does not retain any sign of having been used. Furthermore, the study found, the experience is "depersonalised," because the users found that this material provided no indication of "individuality." Finally, a "partialisation of sensory experience" was identified, because the material precluded any aesthetic expectations that go beyond its pure functionality. It was thought to lack "that extra aesthetic something" which is akin to atmosphere."[1]
At the time, these findings were valid and consensual. However, given that the material's cultural setting has progressed, we have come to the astonishing conclusion that the observations of the time must now be interpreted in a positive light. Resopal, this material without obvious sensory qualities, is now up-to-the-minute precisely because, with its artificiality, agelessness, and neutrality, it is beyond the reach of differentiated sensory experience and qualification. Today, it truly does have an "extra aesthetic something" to offer – in the apparent reproduction of that which it was once accused of lacking.
It is still the same soberly practical, durable, uniform, even, synthetic, hard surface material, once proven to be objective, modern and functionalist with no illusions, as the critical sensualists confirmed. It now adopts such a consistently post-modern attitude that it slots with ease into a transsensory design ethos, the production of virtual experience, appeals to contemporary needs and, in the form of perfected material imitations, is under deep suspicion of having its finger on the pulse of the *Zeitgeist*. Yesterday's shortcomings have mutated into today's virtues. Contemporary culture is concerned with what is not real and tangible, with the removal of all things and surfaces from any historical context, with that which resides permanently in artificial worlds. The Resopal laminate has benefited from this development, indeed it is the obvious choice for something that can make it possible to be drawn into another reality where it is least expected: in the tangible ordinariness of a kitchen countertop.

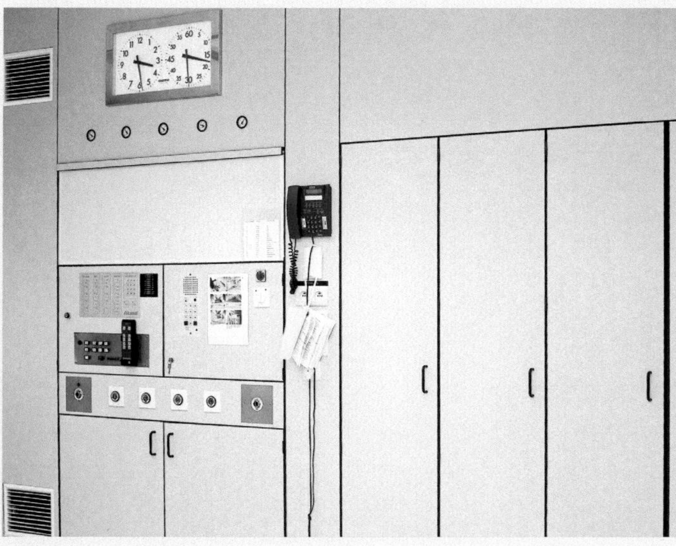

Laborausstattung, chemisch-resistente Flächen mit guter Dekontaminier-
barkeit, 2003
Laboratory fittings: surfaces that are resistant to chemicals and easy to
decontaminate, 2003

ser Entwicklung, sie bietet sich an, den Sog in die andere Realität auch
dort erlebbar zu machen, wo man das nicht erwartet: am handfest Alltäglichen eines Küchen-Arbeitsplatzes.

Rückblickend darf man behaupten, dass das Resopal als zukunftsträchtiges Material zu früh erfunden wurde und sich in seiner zunächst rigi-
den Funktionalität und Sterilität erschöpfen musste, um heute wieder
angesagt zu sein, indem es ein aktuelles ästhetisches Funktionsspekt-
rum entfaltet. Insofern hat dieses spröde Material seine eigene, bewegte, ja durchaus überraschende und überzeugende Geschichte, der man
Gerechtigkeit widerfahren lassen muss, ohne bekennender Resopal-Fan
sein zu müssen.

Man erinnert sich noch an die weißen, hellgrauen oder leicht getönten
Flächen oder an solche mit zarten, abstrakten Gittermustern als Beispiele ästhetischer Ereignislosigkeit. Die heute angebotene Palette der »Plain
Colours« thematisiert farbige Belebungseffekte jedoch neu und liefert
sicherheitshalber deren Interpretation mit: »Reines Weiß ist immateri-
ell und gewissermaßen unsichtbar. (...) Wir haben deshalb schon immer
nicht nur Weiß, sondern eine ausgedehnte Reihe von Off-Whites, von
farbigen Weißnuancen in unserer Kollektion, aus denen pudrige leichte Materialität aufscheint, ebenso schemenhaft transparent wie neblig
beschlagen. Unser neues *Grau 0120 Transition* vermittelt reine Transparenz. Ein reines, unverfärbtes, lichtdurchdrungenes Grau, moderne mate-
rielle Leichtigkeit«[2]

Liest man den Subtext dieser Beschreibung, geht es um die Beseitigung
eines Mangels. Die Homogenisierung von Oberflächen durch immer gleiche Konsistenz der Beschichtung und ihre Einfarbigkeit hatte nicht nur den
Tastsinn enttäuscht, sondern auch dem Auge keinen Anlass gegeben, aktiv
zu werden. Der muss geschaffen werden, notfalls auf dem Wege verbaler
Suggestion. Man darf den Text daher als Gebrauchsanweisung verstehen:
Der Nutzer und Betrachter soll bis ins Feinstofflich-Atmosphärische führende Farbsensationen erleben und das langweilige Material vergessen,
aus dem die wahrgenommene Oberfläche tatsächlich besteht.

Nun ist Resopal, Inbegriff der Nüchternheit, gewiss keine halluzinoge-
ne Substanz. Im experimentellen Design der achtziger Jahre war dieser
Werkstoff gerade wegen seiner spröden Künstlichkeit bei Entwerfern

In retrospect we can say that Resopal as a material with a future was
invented too soon and had to first of all exhaust its potential for rigid
functionality and sterility in order to once more become the *dernier cri*
by developing a contemporary spectrum of aesthetic functions. In that
sense, this austere material has its own eventful, thoroughly surprising
and convincing history, which you can acknowledge without having to
be a devoted fan.

We can still recall white, light-grey or slightly tinted surfaces, or others
patterned with delicate abstract grids as examples of a total lack of aesthetic appeal. But the range of "Plain Colours" on offer today puts colour effects in a new light and, to be on the safe side, provides an explicit
interpretation: "Pure white is immaterial and in a certain sense invisible. (...) Our collections have therefore always included not just white
but an extensive range of off-whites and shades of white with just a
hint of colour, from which a powdery light materiality« emanates that is
shadowy, transparent and misty. Our new *Grey 0120 Transition* conveys
pure transparency: a pure, undyed grey permeated with light, a modern
material lightness."[2]

If we read the subtext of this description it becomes clear that it is about
eliminating a defect: the homogenisation of surfaces through the con-
stant uniformity of the coating and its mono-colour not only disappoint-
ed the tactile sense but gave the eye no incentive to get involved. That
incentive has to be created, if necessary using verbal suggestion techniques. The text can thus be seen as an instruction leaflet: the user and
observer are meant to experience colour sensations that transport them
to ethereal planes, forgetting the mundane material which the surface
perceived is actually made of.

Now Resopal, the epitome of sobriety, is definitely not a hallucinogenic
substance. In the period of experimental design in the 1980s, this material was popular with designers precisely because of its detached artificiality. When attempting to describe the material as possessing sensory
qualities we can either speak of a variation on the story of the emperor's
new clothes or of a perceptual transference of the problem of animat-
ing the surface to the user. The user is invited to get his or her imagination working. Fantasies are called upon to make up for the lack of stim-

Riverside von Resopal, »das übernatürliche Material«, 2004
Riverside by Resopal, "the supernatural material," 2004

beliebt. Entweder darf man beim Versuch, das Material mit sinnlichen Eigenschaften ausgestattet zu beschreiben, von einer Variante des Märchens von des Kaisers neuen Kleidern oder von einer wahrnehmungspsychologischen Verlagerung des Problems der Oberflächenbelebung auf den Nutzer sprechen. Er ist aufgefordert, seine Einbildungskraft zu bemühen. Phantasmen sollen die Reizleere füllen, Fehlendes imaginativ ersetzen. Das Farbdesign der »Plain Colours« wird daher mit einer suggestiven Strategie verbunden: »Wir bringen immer mehr Sinnlichkeit auf die Oberfläche«[3].

Im Design von Dekorplatten wird diese Strategie fortgeführt. Neue Kollektionen von Materialoberflächen-Bildern, die durch ihre Echtheitsanmutung überraschen, perfektionieren die Reizung des visuellen und des haptischen Wahrnehmungs- bzw. Vorstellungsvermögens. Ziel ist »gesteigertes Materialerleben an Stelle bloßer Imitation«. Der Nutzer sieht sich darüber hinaus mit einer Apotheose konfrontiert: »Resopal avanciert zum übernatürlichen Material, das über das Naturvorbild hinaus sinnliches Erleben steigert.«[4] In die Praxis übersetzt heißt das: Man soll eine imitierte Materialoberfläche natürlicher als das Ausgangsmaterial empfinden. Dass das Folgen für die Wahrnehmung des Referenzmaterials hat, ist klar. Es wird einerseits entwertet, um nur noch eine Rolle als das im Vergleich Übertroffene zu spielen, andererseits wird es im perfekten Imitat aufgewertet, das ihm seinen »Sinn« rückerstattet: »Unsere edelmatte Oberfläche EM gibt Holz seine Natürlichkeit zurück, die sonst unter gelackten, gewachsten oder geölten Finishes verschwindet.«[5]

An die Stelle der bei den »Plain Colours« noch geforderten halluzinativen Vergegenwärtigung sinnlich aktivierender Farbvaleurs treten nun veritable Sinnestäuschungen durch Verbildlichung eines dem Anschein nach greifbaren Naturmaterials, das in Wahrheit aus bedrucktem Papier unter einer transparenten Kunstharzschicht, dem Overlay, besteht. Inzwischen gibt es Resopal-Küchenarbeitsplatten in täuschend »echtem« Buchenholzanschliff, Marmor oder Granit, von einem »hochabriebfesten Finish geschützt«, wahrscheinlich widerstandsfähiger als jede natürliche Holz- oder Steinplatte, mindestens säurefest und vor allem: ähnlicher als das Naturmaterial sich selbst sein könnte. Es wird visuell, teils auch mit haptischen Effekten von künstlicher Echtheit oder echter Künstlichkeit über-

uli, the imagination is asked to make up for what is lacking. The "Plain Colour" design is therefore linked to a strategy of suggestion: "We are bringing more and more sensuality to the surface."[3]

This strategy continues in the design of decorative panels. New collections of surface patterns that look astonishingly authentic have perfected the stimulation of our visual and tactile senses, our perception and imagination. The aim is a "heightened experience of the material instead of mere imitation." The user is confronted with an apotheosis: "Resopal has advanced to become a supernatural material that heightens sensory experience to a level above that of its model in nature."[4] Translated into practice that means: we are meant to perceive the material as being more natural than the original. That obviously has consequences for our perception of the reference material: on the one hand, it is devalued, relegated to the role of something that has been superseded. On the other hand, it is upgraded in this perfect imitation that gives it back its "meaning:" "Our satin surface finish restores wood to its true naturalness, which otherwise disappears under varnished, waxed or oiled finishes."[5]

The hallucinative imagining of shades of colour that activate the senses, which the "Plain Colours" still require, are now replaced by real hallucinations caused by a seemingly tangible natural material that in reality consists of printed paper beneath a transparent layer of synthetic resin – the overlay. Resopal kitchen counter tops are now available in deceptively "real" blocked beech, marble or granite, protected by a "finish that is highly abrasion resistant," probably more resistant than any natural wood or stone counter top, resistant to acid and, above all, more like the natural material than the natural material could ever be. It is outtrumped visually and sometimes also with tactile effects that are artificially authentic or authentically artificial. While the eye is occupied, the tactile sense – far less susceptible to deception – glides slightly disconcerted along the artificial surfaces and is merely "carried along" by the visual sense. However, the company is already working on tactile and auditory "authenticity."

The design of these surfaces relies on people's ability to retrieve a real experience of a natural material that was formerly acquired using all the

Resopaloberfläche Marble Siena, 2006
Resopal surface Marble Siena, 2006

boten. Während das Auge beschäftigt ist, gleitet der Tastsinn, weil weniger täuschbar, leicht befremdet an den künstlichen Oberflächen ab und wird nur vom Sehen »mitgenommen«. Aber das Unternehmen arbeitet bereits an der haptischen und akustischen »Echtheit«.

Das Design dieser Oberflächen vertraut auf die Abrufbarkeit einer einst mit allen Sinnen am Naturmaterial erworbenen Realerfahrung. Unwillkürlich stellt sich die Frage, was geschehen wird, wenn eines Tages nur noch die Erfahrung von Simulakren möglich ist. Heute weiß man noch, dass da keine wirkliche Holz- oder Steinplatte vor Augen liegt. Sobald aber das Imitat nur noch sich selber zitiert, saugt es alle Wirklichkeit in sich auf. Jean Baudrillard, der französische Theoretiker der Indifferenz, hat schon vor geraumer Zeit festgestellt, dass es zwischen Simulakrum und Realität keinen Unterschied mehr gibt. Er spricht von der »Ordnung des Hyperrealen und der Simulation«, die kaschiert, »dass das Reale nicht mehr das Reale ist«.[6]

Eben dies scheint sich auf den neu gestalteten Benutzeroberflächen von Resopal abzubilden. Nicht dass man auf der Marmorimitatplatte real arbeiten kann, macht sie zur Benutzeroberfläche, sondern dass auf ihr die Austauschbarkeit des »Echten« und des »Falschen« unter Beteiligung des Benutzers zelebriert wird. Der vergisst, dass Resopal teilweise aus Naturmaterial (Papier) und die Trägerplatte auch aus einem solchen (Holz, Hanf usw.) besteht. Er folgt den visuellen Vorschlägen der Erfahrungsspur, die das Imitat eines anderen Materials legt.

Es sei hier erinnert, dass das Simulationsprinzip historische Vorläufer hat. Inneneinrichter des Barock kamen nicht ohne täuschende Nachahmung von Marmoroberflächen aus, handgemalt auf Holz. Der Celluloidkamm aus den Anfängen des Kunststoffzeitalters musste unbedingt wie aus Schildpatt gesägt aussehen. Die Ersatzfunktionen blieben jedoch erkennbar. Heute ist die Imitatkultur auf dem Höchststand ihrer technischen Virtuosität und Reproduzierbarkeit angelangt. Einfärben, Unterlegen, Aufdampfen, Beschichten, Lüstereffekte, Transparenz – alles kein Problem. War früher die Echtheitsprobe über Auge und Hand jederzeit möglich, werden die Resopal-Oberflächen zum Schauplatz der endgültigen Verabschiedung des Wunsches, eine solche Probe zu versuchen. Resopal erweist sich erneut als ein moderner Werkstoff, weil auf seiner

senses. The question spontaneously arises as to what would happen if one day we could only experience the simulacra. Today, we still know that it is not a genuine wood or stone counter that we see in front of us. As soon as the imitation begins to quote only itself it draws all reality into itself. Jean Baudrillard, the theoretician of indifference, established some time ago that there is no longer any difference between simulacrum and reality. He speaks of the "order of the hyperreal and of simulation," which conceals the fact that "what is real is no longer real."[6]

This is precisely what seems to be depicted on the new user interfaces designed by Resopal. It is not the fact that it is possible to genuinely work on the imitation marble counter top that makes it into a user interface but the fact that the interchangeability of "authentic" and "false" is celebrated with the participation of the user. He forgets that Resopal is made partly from natural materials (paper) and that the board beneath it is also made of natural material (wood, hemp, etc.). He follows the visual suggestions of the experiential trail that the imitation of a different material has laid.

We should recall here that the simulation principle has historical precedents. Interior designers of the Baroque period never failed to employ techniques for deceptive imitation of marble surfaces, hand-painted on wood. The celluloid comb from the early days of the age of plastic absolutely had to look as if it had been cut from a piece of tortoiseshell. But the substitute functions remained recognisable. Today, the imitation culture has reached the highpoint of its technical virtuosity and reproducibility. Dyeing, underlaying, vapour depositing, coating, lustre effects, transparency – no problem at all. Whereas in the past the authenticity test was possible at any time – visually or manually – Resopal surfaces have now become the arena for the definitive farewell to the wish to attempt that kind of test. Resopal is once more proving to be a modern material, because the perceptual confusion can be orchestrated subtly on its surface. Other older covering materials cannot do that. Linoleum, for example, in terms of its constituent parts a natural product and also a product that covers surface areas, was the most modern flooring available before the First World War. Today, it is still, or rather again, being used due to its functional properties and to use colour to accentuate floors in interior

Kletterhalle in Massy, Fassade: Digital in Resoplan, 2003
Climbing hall in Massy, façade: digital in Resoplan, 2003

Oberfläche das Verwirrspiel mit der Wahrnehmung dezent inszeniert werden kann. Andere, ältere Bedeckungsmaterialien leisten das nicht. Zum Beispiel Linoleum, von seinen Bestandteilen her ein Naturprodukt, ebenfalls flächendeckend, vor dem ersten Weltkrieg der modernste Fußbodenbelag, den es gab. Es wird heute noch oder wieder wegen seiner funktionalen Eigenschaften und zur farbigen Akzentuierung von Böden in Innenräumen verwendet. Aber es ist eindeutig nicht modernisierbar im Sinne einer Entmaterialisierung. Linoleum bleibt, was es materialgeschichtlich immer gewesen ist – spürbar körperliche, elastische oder brüchig alternde Materie. Es könnte allenfalls von Resopal oberflächlich imitiert werden, zum Beispiel als Laminat-Fußbodenbelag. Originales Linoleum wäre nicht als Träger von Simulakren geeignet. Das bleibt dem jüngeren Resopal exklusiv vorbehalten.

»Resopal flächig eben, das ist klassisch« (Rudolf Schricker) – der Satz gilt noch immer. Nur dass die Fläche sich nicht mehr selbst darstellt wie bei der monochromen Platte, sondern etwas anderes. Am Anfang war das Opak-Undurchdringliche der Fläche, die neutrale Platte. Dann folgte die verbale Initiation imaginierten sinnlichen Erlebens auf farbigen Flächen. Schließlich traten die Materialimitate mit ihren überzeugenden Täuschungseffekten auf. Die Nichtunterscheidbarkeit bzw. die Oberflächenidentität von Imitat und Vorbild ist das Ergebnis. Der Logik des Prinzips der Simulation folgend, kann das aber nur ein Etappenziel sein. Man fragt sich, ob das Festhalten des Unternehmens am Prinzip Platte mit Materialabbildungen auf ihrer Schau- und Benutzerseite einer unbewussten Treue zur funktionalistischen Tradition geschuldet ist. Nach deren Doktrin war eine Platte nichts weiter als eine Platte, die sich auch im Anwendungsfall nur als solche darstellen durfte. Dass sie heute eine Platte aus einem anderen Material vortäuscht, wäre demnach als Zugeständnis an neue Konsumerwartungen zu verstehen, nicht als Maßnahme einer Modernisierung.

Der nächste Schritt ist aber schon unauffällig vorbereitet. So gibt es Resopal-Dekoroberflächen, die Einblick in eine simulierte Materialtiefe geben, zum Beispiel in die mineralische Struktur einer Steinplatte – durch »tiefgründige Optik von Schärfe und Unschärfe, gepaart mit einer wachsglatten, handschmeichelnden Haptik.«[7]

spaces. But it is clearly not possible to modernise it in the sense of dematerialisation. Linoleum remains what in terms of the history of materials it always was: a perceptibly physical material –either elastic or brittle – that ages. Its surface could even be imitated by Resopal, e.g. as laminate flooring. Original linoleum would not be suitable as a substrate for simulacra. That is the exclusive domain of the more recent Resopal.

"Resopal just flat: that is classic" (Rudolf Schricker) – this is still true. Except the surface no longer depicts itself, as was the case with the monochrome laminated board; it now depicts something else. It all began with the opaque surface – the neutral board. The verbal initiation of imaginary sensory experience on coloured surfaces followed. Finally, the material imitations, producing convincing illusions, made their appearance. The result: it is impossible to distinguish between imitation and exemplar or at least between the identical nature of their surfaces. However, following the logic of the simulation principle, that can only be an interim goal. The question arises as to whether the company's holding on to the principle of a board with materials depicted on the side that is visible and used is a kind of unconscious loyalty to the functionalist tradition. According to its doctrine, a laminated board was nothing more than a laminated board, which, when used, was only allowed to present itself as such. The fact that today a laminated board pretends to be made of a different material would, according to that principle, have to be seen as a concession to new consumer expectations, not as a modernisation measure.

However, the next step has already been discreetly prepared. For example, there are Resopal decorative surfaces that give a glimpse into the depth of the material simulated, for example into the mineral structure of a stone slab – through "a depth of field that combines sharp focus and soft focus, coupled with a waxy smooth touch that is pleasant to the hand."[7]

Modern printing techniques can create the illusion that surfaces have been dissolved to make way for pseudo-spatial effects: "Digital printing permits free-match surface finishes in contemporary immateriality. Soft focus is the third dimension of the surface. It allows a sense of three-dimensional space to be created in places where it would otherwise have been merely depicted."[8]

This fake opening-up of the surface to an imaginary space can be seen

BAR Architecten: Verslaafden Hostel in Utrecht, Interieur: Digital in Resopal, 2003
BAR Architecten: Verslaafden Hostel in Utrecht, interior: digital in Resopal, 2003

Moderne Drucktechnik erlaubt illusionistische Auflösungen von Oberflächen zugunsten scheinräumlicher Effekte: »Der Digitaldruck ermöglicht rapportlose Oberflächen in zeitgemäßer Immaterialität. Unschärfe ist die dritte Dimension der Oberfläche. Sie lässt Raum entstehen, dort wo Raum sonst nur abgebildet wäre.«[8] Noch deutlicher wird die fingierte Öffnung der Oberfläche zu einem vorgestellten Raum bei einem Hotelflur in Utrecht. Dort sind Wände, Türen und Decke mit einem illusionistisch-hyperrealen Laubwerkdekor auf Resopal bedeckt, das in den Zwischenräumen der Blätter Einblick in das Darunter gewährt. Die visuelle Oberflächensperre ist hier scheinhaft aufgehoben.

Inzwischen wird das künstlerische Reflexionspotential der Plattenoberfläche entdeckt. Sozusagen außer Konkurrenz zur gängigen industriellen Serienfertigung von Resopal-Plattenmaterial, das sich schon im ästhetischen Spannungsfeld zwischen Simulation und anwendungsbezogener Benutzeroberfläche bewegt, operiert der Designer und Künstler Tom Stark mit digital inszenierten Effekten der Raumbildung auf Oberflächen, deren Benutzer als differenziert wahrnehmender Kunstbetrachter in Frage kommt: »Tom Stark untersucht Flächen unter digitalen Bedingungen. Die Herstellung dieser Flächen wird von ihm als eine produktkulturell-ästhetische Strategie ähnlich moderiert, wie dies in der Musik durch Remix, Sampling und Scratchen bzw. in der digitalen Bildbearbeitung als Morphing bekannt ist. Sein Projekt beschreibt und relativiert damit einen Entmaterialisierungsprozess, einen von der digitalen Technologie auf allen Ebenen forcierten Übergang vom Materiellen ins Immaterielle, kaum noch Sichtbare. Die Pole von Original und Kopie, Bild und Objekt, Oberfläche und Form verschmelzen und changieren in einer Strategie der Interferenz.«[9]

Digitale Kopien der Kopie einer Oberfläche, sogenannte Vectogramme, die über CNC-Fräsmaschinen in Gestalt dreidimensional-reliefartiger Strukturen dem Plattenmaterial aufgebracht werden, regen zum Nachdenken über das Verhältnis von Original und Nachahmung, Fläche und Räumlichkeitsillusion an. Die künstlerischen Umsetzungen sind hochkomplex und so elaboriert, dass als Rezipient wohl kaum jemand in Frage kommt, der seine Küche mit Resopal verschönern will.

Kurz gesagt: Die Resopalplatte, das historische Banalprodukt schlechthin, erfährt heute ihre künstlerische Nobilitierung als Bildträgeroberflä-

very clearly in the case of the corridor of a rehabilitation centre for drug users in Utrecht. Here the walls, doors and ceiling are covered with an illusionary, hyper-real foliage decor on Resopal, which gives glimpses between the leaves into the space behind. The visual barrier created by a surface appears to have been removed.

The board surface's potential for artistic reflection has now been discovered. Designer and artist Tom Stark – operating "out of competition" as it were from the usual industrial mass production of Resopal laminates, which itself moves in the aesthetic field of tension between simulation and application-based user interface – works with digitally created 3-D effects on surfaces, considering their users viewers of art with differing perceptions: "Tom Stark studies surfaces under digital conditions. He moderates the manufacture of these surfaces using a strategy of product aesthetics that can be compared to the way techniques of remix, sampling and scratching are used in music, or morphing in digital imaging. His project thus describes and relativises a process of dematerialisation, a transition forced at all levels by digital technology from the material to the immaterial, the barely even visible. The separate poles that are original and copy, image and object, surface and form are now merging and changing within a strategy of interference."[9]

Digital copies of the copy of a surface, so-called vectograms, which are applied to the boards in the form of three-dimensional, relief-like structures using CNC milling machines, give pause for reflection on the relationship between original and imitation, the two-dimensional surface and the illusion of three-dimensionality. The artistic renderings are highly complex and so elaborate and sophisticated that as a recipient there is probably scarcely a single laminate customer who wants to beautify his or her kitchen with Resopal.

In brief: the Resopal laminated board, historically the banal product *per se*, is currently enjoying its artistic ennoblement as a substrate for images, on which the digital dematerialisation principle is reflected in a highly artificial manner. The Resopal laminated board as a work of art on the wall – who would have thought that possible even a few years ago? But in terms of market considerations, there are other kinds of surface design with greater general appeal.

Tom Stark: Rohacell, Zellstruktur, CNC-Fräsung, 2005
Tom Stark: Rohacell, cell structure, CNC-milled, 2005

che, auf der das digital-immaterialisierende Prinzip hochartifiziell reflektiert wird. Die Resopalplatte als Kunstwerk an der Wand – wer hätte das noch vor wenigen Jahren gedacht. Marktbezogen bieten sich aber publikumswirksamere Oberflächengestaltungen an.

Heute erwartet man, dass die Auflösung der Oberfläche in tiefenräumliche Illusion weitergeht, in Vollendung der Geschichte des Verschwindens des Realen zugunsten des Irrealen. Man erwartet, dass die feste, abschließende Platte blickdurchlässig wird für dahinter liegende Räume. Was mit einigen materialimitierenden Oberflächen begonnen hat, könnte weiterentwickelt werden, zumindest im Gedankenexperiment. Das Unternehmen fordert schließlich dazu auf: »Schicken Sie uns Ihre Vorlage und wir entwickeln Ihnen Ihr Design.«[10] So darf sich der Designhistoriker ausnahmsweise in der Rolle des Propheten versuchen: Da ein Oberflächendekor nicht nur Hypermaterialität erzeugen, sondern auch in einen virtuellen Raum Einlass gewähren kann, um die banale Platte in ein ebenso fiktives wie multisensuelles Ereignis zu verwandeln, wäre ein *joint venture* zwischen Schichtstofftechnologie und den neuen elektronischen Medien vorstellbar. An Stelle der Dekorpapiere könnten dünne Bildschirmfolien eingelassen werden. Auf ihnen würde zum Beispiel ein leichter Luftzug die Blätter auf der Laubtapete bewegen, vielleicht könnte man gar ihr Rascheln hören. Oder Plattenwände würden sich zu imaginären Räumen öffnen, zu Landschaften, in denen Getreidefelder wogen, Meereswellen sich an Stränden brechen oder weiße Wölkchen vor blauem Himmel treiben. Die virtuelle Entstofflichung der Resopalplatte wäre perfekt. Nicht nur Material, auch Raum, Bewegung und Geräusch fänden sich im Imitat gespiegelt. Nur wer sich verführen ließe, einen dieser Illusionsräume zu betreten, sähe sich schmerzhaft daran erinnert, dass die Platte als das real Undurchdringliche immer noch existiert. Doch wäre der Spielraum entscheidend vergrößert, in dessen Grenzen sich Wahrnehmung und Einbildung bewegen dürfen.

Das klingt utopisch, wäre aber technisch ebenso machbar wie ästhetisch überzeugend, denn die Verwandlung der schlichten Kunstharz-Pressstoffplatte in eine Illusionsmaschine hat ja längst begonnen. Einer immanenten Gesetzmäßigkeit der kulturellen Entwicklung folgend ist sie den Weg vom materiell bestimmten, aber schon tendenziell entsinn-

Today we expect the dissolution of the surface into a three-dimensional illusion to go further, completing the history of the disappearance of what is real to make way for the unreal. We expect the solid board that provides closure to become transparent, allowing us to see spaces "behind it". What began with a handful of surfaces imitating materials could be further developed, at least as a thought experiment. The company after all issues an invitation: "Send us your design and we will develop it for you."[10] Thus the design historian can as an exception try his hand at the role of prophet: Since a decorative finish can produce not only hyper-materiality but can also allow us to gain admission into a virtual three-dimensional space, in order to transform the banal laminated board into an event that is as fictitious as it is multi-sensual, a joint venture between laminating technology and the new electronic media is conceivable. Instead of decor papers, flat-panel monitors in the form of ultra-thin foils could be incorporated. Here, for example, a gentle breeze would move the leaves on the foliage wallpaper, perhaps we would even be able to hear the leaves rustling. Or laminated wall boards would open up into imaginary rooms, landscapes in which cornfields swayed in the breeze, waves break on the ocean's shores or little white clouds dance across a blue sky. The virtual dematerialisation of the Resopal laminated board would be perfect. Not only material, but also space, movement and sound would find themselves reflected in imitations. But, just one thing: anyone who allowed themselves to be seduced into entering one of these illusionary spaces, would be painfully reminded that the board does still exist as something real and impenetrable. Yet the range within which perception and imagination can roam would be decisively enlarged.

That sounds utopian, but it is both technically feasible and aesthetically convincing; because the transformation of the simple laminated board into an illusion machine began some time ago. Following the inherent inner laws of cultural evolution, it followed a particular path, starting with semi-finished duroplastic products, which did have material definition yet were already tending towards the de-sensualized, then on to dematerialised surfaces with imagined stimuli opening up into the illusion of the three-dimensional space and thus virtual self-dissolution.

Vectogramm in Resopal, »die dynamische Oberfläche«, 2003
Vectogram in Resopal, "the dynamic surface," 2003

lichten Duroplast-Halbzeug über immaterialisierte, mit imaginativen Reizen ausgestattete Oberflächen zur Öffnung in die Raumillusion und damit zur virtuellen Selbstauflösung gegangen. Das Produkt bewegt sich auf diesem Weg. Bei einer neuen Kollektion mit Strukturen farbiger Querstreifen ist die Rede vom »gesteigerten Raumerlebnis« und einer Bewegung, die »begrenzende Wände durchdringt«.[11] Die autosuggestive Interpretation des Dekors wirkt wie ein Wunsch, der denkbaren Produktvarianten vorauseilt.

Die Oberfläche der Platte, einst die Kargheit des weißen oder hellgrauen Nichts repräsentierend, hat sich mit Bildern überzogen, die etwas nur im Zitat Anwesendes so zwingend darstellen, dass es als real wahrgenommen werden kann. Der Sog des Immateriellen und der Simulation, untrügliche Anzeichen einer kulturellen Umorientierung, hat die Harzpapier-Pressstoffplatte erfasst, um sie von der Last ihrer materialen Geschichte zu befreien. Würden die Bilder auf ihrer Oberfläche in Bewegung geraten und sich imaginäre Räume hinter ihnen erschließen, schiene ihre Festigkeit aufgehoben. Real bliebe die Platte, was sie war und immer noch ist – eine harte Schicht, die etwas verdeckt. Sie durchlässig zu machen für dahinter liegende Raum-Ereignisse, wäre die konsequente Fortsetzung des Spiels mit der Wahrnehmung, zu dem die Materialbildplatte heute schon animiert.

So hat sich die Resopalplatte als modernisierungsfähiger Entwurf bis hin zu ersten Produktvarianten mit scheinbar »geöffneter« Oberfläche behaupten können. Real ist die Dialektik zwischen Absperrfunktion und tiefenräumlicher Illusion jedoch nicht aufhebbar. Das Produkt bleibt einerseits seiner Geschichte verbunden, andererseits ist es schon bis zur raumöffnenden All-over-Tapete gediehen. Darüber hinaus könnten innovative Bildeinlagerungstechnologien des telematisch-digitalen Zeitalters, die mit der traditionellen Hartpapierpresstechnik kompatibel sind, zu neuen illusionären Oberflächenreizen führen. Eine bildraumillusionistische Beschichtung erscheint ebenso machbar wie die filmisch bewegte und die geräuschaktive Oberfläche. Reale Enge würde so in virtuelle Weite aufgelöst, das totale Als-Ob zur alltagskulturellen Norm erklärt.

Dennoch bliebe die Platte, was sie immer war – eine befestigte Grenze, die nur der Einbildung oder den getäuschten Sinnen offen erscheint.

The product is following this trail. In a new collection of coloured horizontal strip structures, the idea of "heightened experience of space" is cited along with a movement that "penetrates the limits of walls."[11] The auto-suggestive interpretation of the pattern acts like a wish that anticipates conceivable product variations.

The surface of the laminated board, once representing the starkness of white or pale grey nothingness, has been covered with images that depict something that is present only as a quotation, but they depict it so convincingly that it can be perceived as being real. The draw of the immaterial and the simulation, unmistakeable signs of a cultural re-orientation, has caught the laminated board in order to free it from the burden of its material history. If the images on its surface started to move and imaginary spaces behind them were opened up, its solidity would seem to be momentarily suspended. However, in reality the board would remain what it always has been – a hard layer that covers something. Making it permeable to three-dimensional events "behind" it, would be the logical continuation of the play with perception which today's image substrate inspires in us.

Thus, the Resopal laminated board has been able to assert itself as a design that can perfectly well be modernised, including the first product variations with apparently "opened-up" surfaces. In reality, however, the dialectic between barrier function and the illusion of three-dimensional space cannot be eliminated. The product remains, on the one hand, firmly rooted in its history, while, on the other hand, it has already blossomed into all-over wallpaper that opens up space. Furthermore innovative imaging technologies of the telematic, digital age that are compatible with the traditional laminating technology could produce new illusionary surface stimuli. A coating that creates the illusion of three-dimensionality seems just as feasible as a surface animated by film or sensitive to sound. Real constraint would thus be dissolved in virtual expanse, the total "as if" experience would be declared a normal part of daily life.

Yet the board would remain what it has always been – a fixed boundary that only appears open to our imagination or deceived senses.

Gerd Ohlhauser

DIE MACHER DER OBERFLÄCHE
THE CREATORS OF THE SURFACE

War Resopal ursprünglich weit mehr als Laminat (Günter Lattermann), ist es heute genau genommen nicht einmal mehr das. Das 1930 patentierte Nachfolgeprodukt des dunkelbraunen Bakelits, das endlich auch hellfarbig eingefärbt werden konnte, stand anfangs für Kunststoff-Pressmassen, aus denen bunte Haushaltsgegenstände gefertigt wurden. Erst mit dem Erfolg der damit imprägnierten Platte in den fünfziger Jahren wurde Resopal der Name für die Laminatoberfläche. Doch in dem, was die 1971 nach ihrem bekanntesten Produkt umbenannte Firma Resopal heute unter diesem Namen verkauft, steckt schon lange kein Resopal mehr.

Das Resopal genannte Laminat besteht heute aus mit Phenol- und Melaminharz imprägnierten Papierbahnen, die mit großer Hitze unter hohem Druck verpresst werden. Die verwendeten Kunstharze sind nicht mehr Resopal, und die charakteristischen Oberflächendekore kommen nicht mehr aus dem Hause Resopal. Grundstoffe und Leistungen sind zugekauft. Die spezifische Qualität der Resopal-Platte besteht nur noch in dem, was die Firma Resopal daraus macht. Da die gleichen Grundstoffe und Leistungen aber auch von jedem anderen Hersteller eingesetzt werden, unterscheiden sich die Oberflächen kaum noch voneinander. Das erklärt, warum der Name Resopal im Ursprungsland Deutschland zum Synonym für alle Oberflächen dieser Art wurde. Resopal hat der Oberflächen-Branche den Weg bereitet, doch die Weiterentwicklung des Produktes betreiben vor allem ihre hochspezialisierten Zulieferer.

Am augenscheinlichsten wird dies am Dekorpapier, das letztendlich die Ästhetik der Oberfläche ausmacht. Aus der 1928 in Berlin mit der Anmeldung des Dekortiefdruckpatents gestarteten Fima **Masa**, die ihrem Auftraggeber Resopal nach dem Zweiten Weltkrieg noch an den neuen Standort folgte, ist mit **Interprint** (1969), **Süddekor** (1974), **Schattdecor** (1985) und in jüngster Zeit **Bausch Decor** inzwischen eine weltweit agierende Branche hervorgegangen, die ihren Kunden längst die Dekorthemen und -trends vorgibt. Weltweit wird heute das Oberflächendesign für Möbel, Fußböden, Deckenpaneele und den Innenausbau fast ausschließlich von diesen Firmen aus Deutschland gemacht – paradoxerweise dem Land mit den mächtigsten Kulturhütern (Werkbund, Bauhaus, Rat für Formgebung, Hochschule für Gestaltung Ulm), denen Imitation und Oberflächendekor immer schon als »Verbrechen« (Adolf Loos) galt.

Auch bei den »inneren Werten« hat der Siegeszug der Oberfläche zur Ausdifferenzierung einer eigenständigen Zulieferindustrie geführt. Die Bakelite AG, Nachfolgerin der von dem Erfinder des Bakelits Leo Hendrik Baekeland und dem Unternehmer Julius Rütgers 1910 in Berlin gegründeten Bakelite Gesellschaft, baut ihre fast hundertjährige Erfahrung in der Phenolharztechnologie nach der kürzlichen Übernahme durch **Hexion Specialty Chemicals** verstärkt aus. Die stetige Optimierung des Melamins hinsichtlich Härte und Widerstandsfähigkeit einerseits sowie Fließverhalten und Transparenz andererseits betreibt seit langem die österreichische **AMI Agrolinz Melamine International**, die aus der 1939 gegründeten »Stickstoffwerke Ostmark AG« hervorgegangen ist. Eines der saugfähigsten, aber gleichzeitig genügsamsten Kernpapiere kommt heute von der auf das Jahr 1880 zurückgehenden amerikanischen Firmengruppe **MeadWestvaco**. Auf dünne Overlays (transparente Deckschicht), die gleichzeitig höchstes Sättigungsvermögen besitzen, ist die seit 1998 zum amerikanischen Glatfelter-Konzern gehörende Schwarzwälder Papiermühle **Schoeller & Hoesch** spezialisiert. Die gleichmäßig ansaugenden Farbpapiere werden von **Arjowiggins** in der seit 1492 bestehenden französischen Papiermühle in Arches hergestellt, die neben anderen berühmt gewordenen Papieren 1869 zusammen mit dem Maler Dominique Ingres das Ingres-Papier entwickelte.

Alle diese Zulieferer stellen ihren Beitrag zur Ästhetik der Oberfläche auf den folgenden Seiten dar.

Although Resopal was originally far more than just a laminate (Günter Lattermann), today, to be precise, it is not even that. The successor product to the dark-brown Bakelite, which was patented in 1930 and had the advantage that finally it could also be dyed in light colours, was initially a plastic moulding compound that was used to manufacture brightly coloured household goods. It was not until sheets impregnated with it became such a success in the fifties that Resopal became the generic name for laminates per se. However, the product made by Resopal, as the company itself has been known since 1971 when it adopted the name of its most famous product, has not actually contained Resopal for a long time.

The laminate known as Resopal now consists of sheets of paper impregnated with phenolic and melamine resins and fused together under high pressure at a high temperature. The synthetic resins used are no longer Resopal and the typical surface designs are no longer created within the company. Services and raw materials are acquired. The specific quality of Resopal sheets consists solely in what the company makes of it. However, since other manufacturers use the same services and raw materials, there is very little difference in these surfaces. That explains why Resopal became the synonym for all laminates of this kind in Germany, the country of its origin. Resopal laid the foundation for the laminate industry, but it is primarily their highly specialised suppliers who refine the product today.

This is most evident in the case of decor paper, which is what ultimately determines the aesthetics of the surface. Out of the **Masa** company, which took up business in Berlin in 1928 after having filed the patent for decor paper gravure printing with **Interprint** (1969), **Süddekor** (1974), **Schattdecor** (1985) and recently **Bausch Decor**, a globally active branch emerged which guides its customers in decor themes and trends. Throughout the world the designs for furniture, flooring, ceiling panels and interior surfaces are now being produced almost exclusively by these companies from Germany – ironically the country with the most powerful "guardians of culture" (Werkbund, Bauhaus, Rat für Formgebung, Hochschule für Gestaltung in Ulm), who always considered imitation and decorative surfaces to be a "crime" (Adolf Loos).

Also regarding the "inner values," the triumph of the surface led to the formation of an autonomous supplier industry. Bakelite AG, the successor to Bakelite Gesellschaft founded in Berlin in 1910 by the inventor of Bakelite Leo Hendrik Baekeland, with the entrepreneur Julius Rütgers, and just recently integrated into **Hexion Specialty Chemicals**, continues to build on its almost 100 years of experience with phenolic resin technology. For a long time now, the Austrian company **AMI Agrolinz Melamine International**, the successor to "Stickstoffwerke Ostmark AG," established in 1939, has been working on the steady optimisation of melamine in terms of hardness and resistance on the one hand and flow behaviour and transparency on the other. One of the most absorbent yet economical core papers on the market today comes from **MeadWestvaco**, an American group whose origins go back to 1880. **Schoeller & Hoesch**, a paper mill based in the Black Forest that has been part of the American Glatfelter corporation since 1998, specialises in thin overlays (transparent top layers of paper) of highest saturation capacity. Coloured papers that absorb evenly are manufactured by **Arjowiggins** in the French town of Arches in a paper mill that dates back to 1492. The mill developed many famous papers, including in 1869 the Ingres paper in conjunction with the painter Dominique Ingres.

The contribution to the aesthetics of the surface made by these suppliers is illustrated in the following pages.

Je weiter man zurückblicken kann, desto weiter kann man vorausschauen.

The farther backward you can look, the farther forward you are likely to see.

Winston Spencer Churchill

Es erfüllt uns mit Stolz, zurückzublicken auf 75 Jahre gemeinsamer Dekorentwicklung,
die Kulturgeschichte geschrieben hat. MASA-Decor GmbH, Dreieich
We look back with pride on the 75 years of mutual decor development that has
helped write cultural history. MASA-Decor GmbH, Dreiech.

INTERPRINT – DER AUFSTIEG DER DEKORATIVEN OBERFLÄCHE
INTERPRINT – THE RISE OF THE DECORATIVE SURFACE

Interprint ist das erste Dekordruck-Unternehmen, das seine Entstehung dem Aufstieg der dekorativen Oberfläche verdankt. Es wird 1969 in Arnsberg von jenem Paul Wrede gegründet, der dort bereits 1956 das Resopal-Konkurrenzprodukt Duropal auf den Markt brachte. Die Entstehung aus dem Hochdrucklaminat Duropal ist zwar bis heute prägend, dennoch ist das Unternehmen von Anfang an auf den internationalen Markt ausgerichtet. Mit Standorten in Deutschland, Italien, Polen, Russland, USA, Malaysia und China zählt Interprint zu den weltweit führenden Dekordruckern. Seit 2006 verfügt Interprint als erstes Dekordruck-Unternehmen über eine eigene Lasergravur, die nicht nur einen Qualitätssprung garantiert, sondern ganz neue Möglichkeiten eröffnet. Aktuelles Beispiel ist das für Resopal entwickelte Dekor „Soft Wear", dessen Gleichmäßigkeit im Farbverlauf bis dato nicht machbar war.

Interprint is the first decor paper printer that owes its formation to the rise of the decorative surface. It was founded in Arnsberg, Germany in 1969 by the same Paul Wrede who already had brought Resopal's rival product Duropal to the market in 1956. Although it emerged from the high-pressure laminate Duropal, the company has focused on the international market right from the beginning. With locations in Germany, Italy, Poland, Russia, USA, Malaysia and China, Interprint is one of the worldwide leading decor printers. Since 2006 Interprint is the first decor printing company with a laser engraving system. This guarantees not only a leap forward in quality, but also opens up entirely new possibilities. One of the first outcomes is the decor "Soft Wear," developed especially for Resopal, whose perfect colour gradation has not been feasible until now.

Dekor „Soft Wear" für Resopal
Mit der Lasergravur beginnt die Auflösung der Oberfläche.
Decor "Soft Wear" for Resopal
laser engraving marks the beginning of the surface's dissolution

www.interprint.com

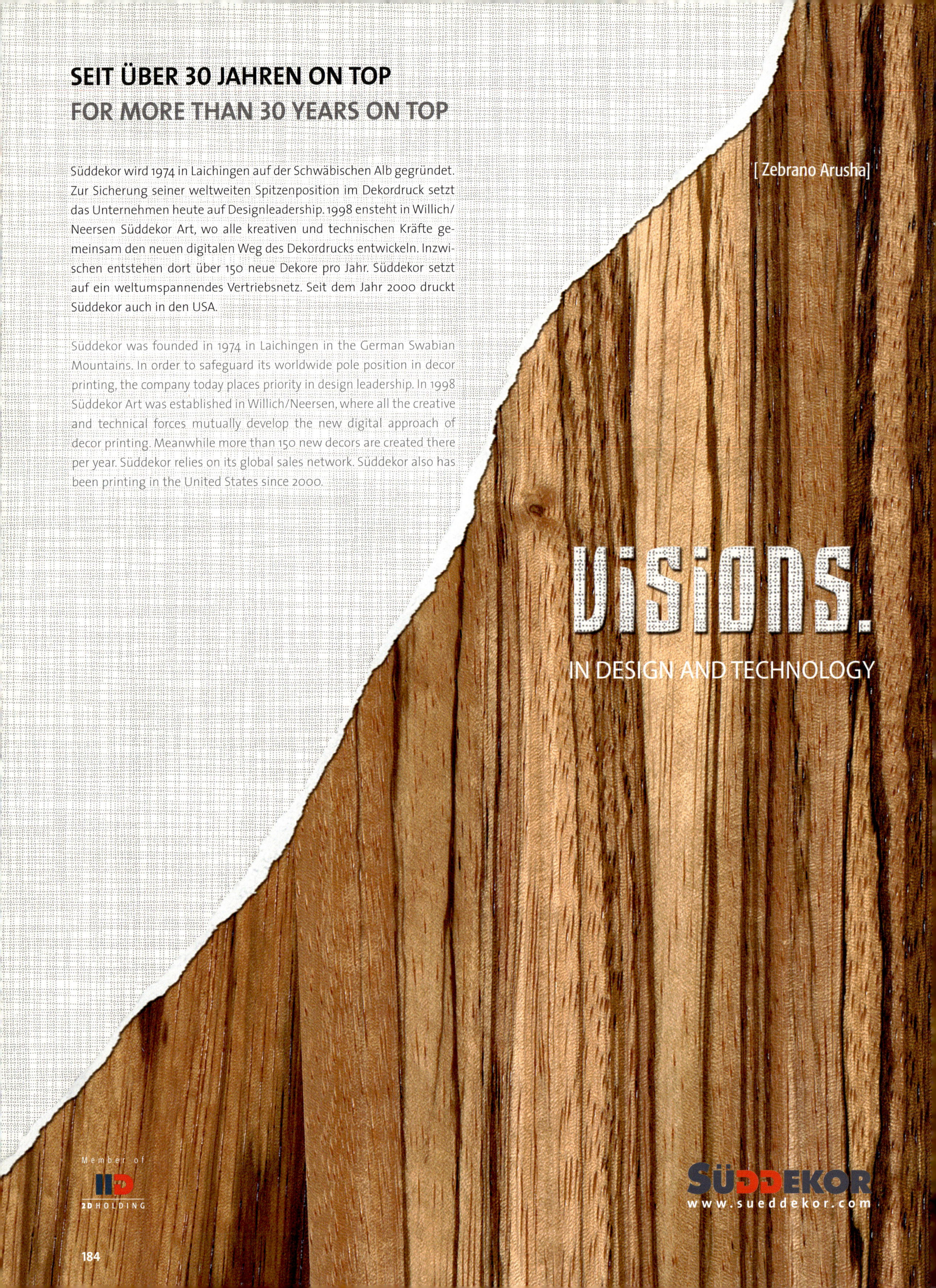

SEIT ÜBER 30 JAHREN ON TOP
FOR MORE THAN 30 YEARS ON TOP

Süddekor wird 1974 in Laichingen auf der Schwäbischen Alb gegründet. Zur Sicherung seiner weltweiten Spitzenposition im Dekordruck setzt das Unternehmen heute auf Designleadership. 1998 ensteht in Willich/Neersen Süddekor Art, wo alle kreativen und technischen Kräfte gemeinsam den neuen digitalen Weg des Dekordrucks entwickeln. Inzwischen entstehen dort über 150 neue Dekore pro Jahr. Süddekor setzt auf ein weltumspannendes Vertriebsnetz. Seit dem Jahr 2000 druckt Süddekor auch in den USA.

Süddekor was founded in 1974 in Laichingen in the German Swabian Mountains. In order to safeguard its worldwide pole position in decor printing, the company today places priority in design leadership. In 1998 Süddekor Art was established in Willich/Neersen, where all the creative and technical forces mutually develop the new digital approach of decor printing. Meanwhile more than 150 new decors are created there per year. Süddekor relies on its global sales network. Süddekor also has been printing in the United States since 2000.

[Zebrano Arusha]

Visions.
IN DESIGN AND TECHNOLOGY

Member of
IID
IID HOLDING

SÜDDEKOR
www.sueddekor.com

20 JAHRE BESTSELLER
20 YEARS OF BEST-SELLING DECORS

Das von Walter Schatt 1985 gegründete Unternehmen Schattdecor steigt bis 1998 in nur dreizehn Jahren zum Weltmarktführer für bedruckte Dekorpapiere auf. Schattdecor ist das Ergebnis der akkumulierten Erfahrung, die Walter Schatt in der Dekordruckbranche gesammelt hat. Die Unternehmensgruppe beschäftigt heute weit mehr als 1300 Mitarbeiter am Stammsitz in Thansau bei Rosenheim, in Italien, Polen, Russland, Brasilien und China. Mit ihrer Beteiligung am Schweizer Druckfarbenhersteller Arcolor, dem eigenen Druckmaschinenbau-Unternehmen Rotodecor in Bad Salzuflen und eigener Dekorpapierproduktion mit einem chinesischen Partner besitzt die Familien-AG alle notwendigen Ressourcen für die Versorgung des Weltmarktes mit bedruckten Dekorpapieren made by Schattdecor.

Von Beginn an setzen Schattdekore Standards und gelten als eigenständige Originale jenseits des Imitats. Heute verfügt die zentrale Dekorabteilung in Thansau über nahezu 1000 Dekore in mehr als 10.000 verschiedenen Farbstellungen.

The printed decor paper manufacturer Schattdecor was established in 1985. By 1998, a mere 13 years later, it had reached the position of world market leader thanks to the decor paper printing experience of its founder Walter Schatt. Today the Schattdecor Group has over 1,300 employees working at its headquarters in Thansau, near Rosenheim, Germany, and at printing plants in Italy, Poland, Russia, Brazil and China. Thanks to a joint venture with a Chinese partner in decor paper production, a participating interest in the Swiss printing ink producer Arcolor and outright ownership of the printing press manufacturer Rotodecor in Bad Salzuflen, Germany, the family-owned public limited company is able to supply the world market with printed decor paper that can truly claim to be "made by Schattdecor."

Schatt decors have been setting standards from the very outset and are regarded less as imitations and more as originals in their own right. The Group's main design department in Thansau can offer a choice of some 1,000 decors in over 10,000 colourways.

schattdecor

1985 Die geplankte Bavaria Buche beendet endgültig das dunkle Zeitalter der geblumten Eiche. Bavaria Beech, a planked version of the woodgrain, sees out the era of slip match oak.

1987 Der Thansau Ahorn eröffnet die Ära der hellen heimischen Hölzer. Thansau Maple lends pale indigenous wood a new popularity.

1989 Der naturbelassene Kirschbaum begründet die Neue Natürlichkeit der neunziger Jahre. Schattdecor's natural-look cherry marks the start of the New Authenticism of the nineties.

1991 Die Farbe der Roterle wird zum Holzton des Jahrzehnts. The warm hue of Schattdecor's red alder becomes the woodgrain colour of the decade.

1993 Die neue Samerberg Buche macht die Buche endgültig zum aktuellen Designerholz. Beech becomes a new favourite with designers thanks to Samerberg Beech.

1995 Mit Buche Schiffsboden wird der Laminatboden salonfähig. Schiffsboden Beech makes laminate flooring socially acceptable.

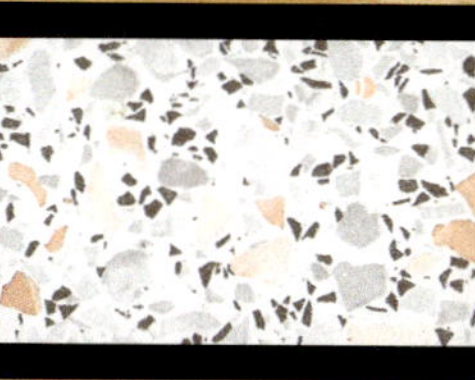

1997 Das Terrazzo-Dekor Ravenna bricht die Vorherrschaft der Granitimitation auf der Küchenarbeitsplatte. Ravenna, a mosaic-look decor, ends the supremacy of imitation granite on kitchen countertops.

1999 Der Feldberg Ahorn revitalisiert den mittlerweile zum Klassiker gewordenen Thansau Ahorn. Feldberg Maple renews the popularity of the longstanding classic Thansau Maple.

2001 Die Sacramento Pine macht Vintage zum Trend. Sacramento Pine sets the vintage trend.

2003 Mit Granito Rojo beginnt eine neue Klassik. A new classic is born in the form of Granito Rojo.

2005 Die Wallis Zwetschge entwickelt sich zum nächsten Bestsellerdekor. Valais Plum emerges as the next best seller.

Ein Hauch von **Ornament**

DEKORENTWICKLUNG VON BAUSCH DECOR

Mit dem Dekor Nocero Ornament – ausgezeichnet vom Innovations Scouting ZOW 2006 – zeigt Bausch Decor, wie gut moderne Formensprache und Ornament harmonieren können. Inspiriert von der „Arts and Crafts"-Bewegung ist die Furnierintarsie als stilisiertes Blütenornament entwickelt worden, bei dem in ein ruhiges Ahorndekor Nussbaumblüten eingesetzt sind. Durch die zurückhaltende Farbigkeit und das edle Design erregt Nocero Ornament Aufsehen ohne dominant zu sein.

Die 1999 im bayrischen Buttenwiesen gegründete Bausch Decor ist eine 100-prozentige Tochtergesellschaft der börsennotierten Surteco AG, einem der weltweit führenden Spezialisten für Oberflächentechnologien. Von Anfang an macht das Unternehmen mit außergewöhnlichen Designs Furore.

A Hint of **Ornament**

DECOR DEVELOPMENT FROM BAUSCH DECOR

With the decor Nocero Ornament—distinguished with a prize by Innovations Scouting ZOW 2006—Bausch Decor proves how well modern design and ornament can harmonize. Inspired by the "Arts and Crafts" movement, the veneer inlay has been designed as a stylized ornamental blossom, with nut-tree blossoms set into a tranquil maple wood decor. Through its muted colouration and refined design, Nocero Ornament arouses attention without being dominant.

Bausch Decor, established in the Bavarian town of Buttenwiesen in 1999, is a 100-percent subsidiary of the listed company Surteco AG, a worldwide leading specialist for surface technologies. The company has been creating a stir with its extraordinary designs since the company's inception.

PHENOLHARZ VON BAKELITE®

Bakelite® – dieser Name steht nicht nur für den weltweit ersten vollsynthetischen Kunststoff, sondern auch für dessen Herstellungsprozess und bis vor kurzem auch für dessen Hersteller.

Der belgische Chemiker Leo Hendrik Baekeland meldet das nach ihm benannte Bakelite®-Patent am 18. Februar 1907 zunächst in den USA an. 1909 überträgt er die Patentrechte für Europa an die deutschen Rütgers-Werke in Erkner bei Berlin, mit denen er 1910 die Bakelite®-Gesellschaft gründet. Diese beginnt als erstes Unternehmen der Welt mit der Großproduktion von vollsynthetischen Kunststoffen. Die Nachfolgegesellschaft, die Bakelite AG, wächst in den folgenden Jahrzehnten zu einem der weltweit führenden Hersteller von Phenol- und Epoxidharz. Heute werden die Bakelite-Produkte unter dem Dach des neu gegründeten Unternehmens Hexion Specialty Chemicals hergestellt, das 2005 im Rahmen einer Fusion von Bakelite AG, Borden Chemical, Resolution Performance Products und Resolution Specialty Materials entstand und damit zum weltgrößten Hersteller duroplastischer Kunststoffe avancierte.

1919 erhält die H. Römmler AG, die heutige Resopal GmbH, die zur selben Zeit das gleiche Verfahren entwickelt hat, ein lizenzfreies Mitbenutzungsrecht des Bakelite®-Patents. Mit dem am 19. Dezember 1930 angemeldeten Resopal-Patent (Resina pallida – lat. für bleiches Harz) modifiziert Römmler das dunkelfarbige Bakelite® zu einem glasklaren Harz, das erstmals auch hell eingefärbt werden kann, und schafft so die Voraussetzung für die später berühmt werdende Resopalplatte. Die Bakelite AG entwickelt die Eigenschaftsprofile seiner Phenolharze ständig weiter und wird nicht nur der größte Zulieferer der Resopal GmbH, die damit ihre Kernpapiere imprägniert, sondern nahezu der gesamten Span-, Faserplatten- und Furniersperrholz-Industrie.

BAKELITE® – DER STOFF DER TAUSEND MÖGLICHKEITEN

Dieser Slogan aus den dreißiger Jahren beschreibt die revolutionäre neue Vielfalt. Denn: Kunststoffe sind Multitalente. Sie sind in der Regel beständiger, haltbarer und leichter als „natürliche Werkstoffe", können in ihren Eigenschaften gezielt beeinflusst und in nahezu jede Form gebracht werden.

Mit Bakelite® entsteht die Technik des Pressformens. Alle Arten von Gebrauchsartikeln, vom Elektrostecker bis zum Aschenbecher, vom Radiogehäuse bis zum Spielzeug, können auf einmal aus einem Guss in großer Stückzahl kostengünstig hergestellt werden. Mit ihren presstechnisch bedingten, charakteristischen Gestaltungsmerkmalen wirken sie gleichzeitig prägend auf den Stil der Zeit. So finden die abgerundeten Ecken und Kanten, die gemaserten Oberflächen und edeldunklen Farben des Art déco hier ihren Ursprung.

PHENOLIC RESINS FROM BAKELITE®

Bakelite® —this name stands not only for the first fully synthetic resin, but also for its manufacturing process, and up to recently for the manufacturer as well.

The Belgian chemist Leo Hendrik Baekeland first applied for the patent on the Bakelite® material named after him on February 18, 1907 in the USA, and in 1909 assigned the European patent rights to the German Rütgers-Werke, with whom he founded the Bakelite® Company in Erkner near Berlin in 1910. This company was the first in the world to initiate industrial production of fully synthetic plastics. In the following decades, the corporation that developed from it, Bakelite AG, grew to become one of the world's leading manufacturers of phenolic and epoxy resins. Today, Bakelite® products are manufactured under the management of the newly founded Hexion Speciality Chemicals Group that emerged in 2005 from the fusion of Bakelite AG, Borden Chemical, Resolution Performance Products and Resolution Speciality Materials and thus advanced to the world's largest manufacturer of thermosets.

In 1919, H. Römmler AG — now Resopal GmbH — which at the very same time developed the same process, was assigned a license-free right to co-exploit the Bakelite® patent. In its Resopal patent (Resina pallida — Latin for "pale resin"), applied for on December 19, 1930, Römmler modified the dark-coloured Bakelite® to obtain a crystal clear resin that could be produced in light colours for the first time and thus setting the stage for the later famous Resopal sheets. Bakelite AG continuously advanced the product profiles of its phenolic resins, not only becoming the main supplier of Resopal GmbH, which uses the resin to impregnate its core papers, but also of the entire chipboard, fibreboard and veneer plywood industry.

BAKELITE® —THE MATERIAL OF A THOUSAND USES

This advertising slogan from the 1930s describes the revolutionary new versatility of the material. Since in fact, plastics are all-round products. They are generally more resistant, durable and lightweight than "natural materials," can be modified to obtain specific properties, and are capable of being made into nearly any shape.

The technique of compression moulding arose with Bakelite®. From that point on, all kinds of consumer articles, from electric plugs to ashtrays, from radio housings on to toys, could be economically produced in a single piece and in large numbers. Because of their typical compression moulding design features, such products simultaneously put their mark on the style of the age. Thus, the rounded corners and edges, the speckled surfaces and elegant dark colours of Art Déco have their origins here.

Dr. Leo Hendrik Baekeland

Baekelands Reaktionsgefäß des ersten Kunststoffs
Baekeland's reaction vessel for the first synthetic resin

Bakelite®-Logo

MELAMIN VON AGROLINZ MELAMINE INTERNATIONAL
MELAMINE FROM AGROLINZ MELAMINE INTERNATIONAL

Mit acht Produktionsanlagen in Deutschland, Italien und Österreich deckt die Agrolinz Melamine International, der weltweit zweitgrößte Melaminproduzent mit Hauptsitz in Linz, nahezu ein Viertel des globalen Melaminbedarfs von 900 000 Tonnen jährlich. Das aus der 1939 gegründeten Stickstoffwerke Ostmark AG hervorgegangene Unternehmen versorgt die „Harzfabrik" von Resopal nicht nur mit Melaminpulver für die Produktion einer der härtesten und unempfindlichsten Oberflächen, sondern auch mit unterschiedlichsten Additiven, mit denen die Resopal-Oberflächen gezielt für die jeweiligen Anforderungen ausgerüstet werden. Jüngste Entwicklung ist HIPERCARE® (High Performance Care), das nachweislich das Ansiedeln und Aufwachsen von Mikroorganismen auf der Oberfläche verhindert. Die Wirkstoffe von HIPERCARE® sind in dem Melaminharz gelöst, mit dem die dekorativen Deckschichten von Resopal durchtränkt sind, und entfalten ihre Wirksamkeit – toxikologisch völlig unbedenklich – beim direkten Kontakt. Die Melaminoberfläche, von jeher eine der hygienstischen Oberflächen, wird mit HYPERCARE® hygieneaktiv.

With its eight production facilities in Germany, Italy and Austria, Agrolinz Melamine International, the world's second largest melamine producer with headquarters in Linz, Austria, covers almost one quarter of the annual global melamine requirement of 900,000 tons. The company, derived from the 1939-established Stickstoffwerke Ostmark AG, supplies Resopal's "resin factory" not only with melamine powder for the production of one of the hardest and most resistant surfaces. It also provides various additives needed to equip Resopal surfaces for individual requirements. The most recent development is HIPERCARE® (High Performance Care) which demonstrably prevents the settling and growth of micro-organisms on the surface. The active agents of HIPERCARE® are dissolved in the melamine resin with which the decorative covering layers of Resopal are impregnated, and unfold their effectiveness—toxicologically completely harmless—in direct contact. The melamine surface, which always has been one of the most hygienic surfaces, becomes hygienically active through HIPERCARE®.

HIPERCARE® von AMI Agrolinz Melamine International macht Resopal hygieneaktiv.
HIPERCARE® of AMI Agrolinz Melamine International makes Resopal hygienically active.

KERNPAPIERE VON MEADWESTVACO
CORE PAPERS FROM MEADWESTVACO

Westvaco, heute MeadWestvaco, gegründet 1880, begann in den vierziger Jahren in Virginia mit der Produktion von Kraftpapieren. Anfang der sechziger Jahre kam es zur technischen Zusammenarbeit mit Resopal, seit 1982 ist MeadWestvaco dessen Hauptlieferant für Kraftpapiere. Diese wurden anstelle der losen Cellulosefasern in den Pressformteilen entwickelt, um plattenförmiges Material herzustellen. Sie bilden den tragenden Kern der Resopalplatte und heißen deshalb auch Kernpapiere. Durch die Imprägnierung mit Phenolharz werden sie feuchtigkeits- und hitzebeständig und verleihen dem Laminat seine hohe Festigkeit und Schlagzähigkeit bei gleichzeitig höchster Maßhaltigkeit. MeadWestwaco-Papiere zeichnen sich durch eine Faserkomposition aus, die optimal auf die Weiterverarbeitung und den Einsatz des Endproduktes abgestimmt ist und maximale Produktivität bei minimaler Harzabsorbtion garantiert. Durch entsprechende Zusätze können sie eingefärbt oder feuerfest gemacht werden. Mit der Verfeinerung seiner Test- und Kontrollmethoden und der Weiterentwicklung der Rohmaterialien arbeitet MeadWestvaco an der stetigen Verbesserung seiner Produkte.

Seit den späten fünfziger Jahren produziert MeadWestvaco auch Overlays, die oberste, den Dekordruck schützende, unsichtbare Schicht des Laminats. Die nur 25 bis 30 g/qm wiegenden Papiere werden aus Fasern hergestellt, die den gleichen Lichtbrechungsindex wie das Melaminharz haben, mit dem sie getränkt werden, und infolgedessen nach der Verpressung transparent erscheinen. Durch die Einlagerung von sehr feinem, weißen Korund konnte MeadWestvaco die Abriebfestigkeit der Oberfläche so weit steigern, dass diese ab 1990 auch fußbodentauglich wurde. Das war der Beginn der beispiellosen Erfolgsgeschichte des Laminatfußbodens, von dem bereits 10 Jahre später 200 Millionen Quadratmeter verkauft wurden – Prognosen gehen für 2006 von 750 Millionen Quadratmetern aus. Danach war es nur konsequent in Overlays auch dekorative Materialen wie holografische, metallisch schimmernde Partikel oder Naturfasern einzubauen, die damit erstmals eine eigenständige Materialität der Oberfläche hervorbrachten.

Westvaco, today MeadWestvaco, founded in 1880, began producing kraft papers in Virginia in the 1940s. In the beginning of the 1960s, a co-operation with Resopal was established, and since 1982 MeadWestvaco is its main supplier for kraft paper. Instead of the loose cellulose fibre for press moulded articles, they were developed in order to produce sheet material. They form the supporting core of the Resopal sheet, and therefore are called core papers. Thanks to the impregnation with phenolic resin they become water- and heat-resistant. They lend the laminate its high resistance, and at the same time highest dimension stability. MeadWestvaco papers are excellent for their fibre composition which is in optimum compliance with further manufacturing cycles and the use of the end product, and warrants maximum productivity with minimum resin absorption. By using the relevant additives, they can be dyed or made fireproof. By refining its test and control methods and further development of the raw materials, MeadWestvaco is steadily improving its products.

Since the late 1950s MeadWestvaco also is producing overlays, the uppermost invisible protective layer of the laminate. The papers weighing only 25 to 30 g/m^2 are made of fibres which have the same light-breaking index (index of refraction) as the melamine resin they are impregnated with, and consequently they become transparent after pressing. By introducing very fine white corundum, MeadWestvaco was able to increase the abrasion resistance of the surface to such an extent that the material also became suitable for flooring as of 1990. This was the beginning of the unparalleled success story of the laminate floor, of which already ten years later, 200 million square metres have been sold. Forecasts for 2006 assume 750 million square metres. Consequently, it was only logical to include also decorative materials like holographic, metallically shining particles or grass and other natural fibres, which thereby produced for the first time a surface materiality of its own right.

MeadWestvaco

Der Aufbau der Resopalplatte vor der Verpressung: Kraftpapiere von MeadWestvaco – der tragende Kern der Resopalplatte.
The build up of the Resopal sheet before the pressing: kraft papers of MeadWestvaco — the supporting core of the Resopal sheet.

HART IM NEHMEN:
TRANSPARENTER OBERFLÄCHENSCHUTZ

SCHOELLER & HOESCH STELLT OVERLAYPAPIERE FÜR UNTERSCHIEDLICHSTE MÖBEL- UND LAMINATVERBUNDSTOFFE HER:

- Hochgefüllte, korundhaltige Overlaypapiere für die Herstellung von Laminatfußböden für alle Abriebklassen von AC 1 bis AC 5 und hochabriebbeständige Beschichtungen für OSB (Oriented Strength Boards) und Sperrholz für die Fahrzeug- und Bauindustrie (Flächenmasse von 18–75 g/m^2).
- Ungefüllte Overlaypapiere für die Herstellung von Flüssig-Overlays für alle Herstellungsverfahren für LPL- und HPL-Einsatz (Flächenmasse 18–35 g/m^2).
- Ungefüllte Overlaypapiere für die Herstellung von Overlay-Melaminfilmen für die Außenanwendung (Flächenmasse 16–30 g/m^2).
- Gefüllte Overlaypapiere mit einem Korundgehalt < 10 % für die Herstellung von HPL, CPL und LPL, für Arbeitsplatten und beschichtete Holzwerkstoffplatten mit hoher Abriebbeanspruchung.
- Gefülltes und ungefülltes Overlaypapier für die Herstellung von Formholz (z. B. Tabletts oder Formholzstühle).
- Naturfarbenes Overlaypapier für Furniere und dunkle Druckdekore.

SCHOELLER & HOESCH OVERLAYPAPIERE
SCHOELLER & HOESCH OVERLAY PAPERS

Schoeller & Hoesch wird 1881 in Gernsbach im Schwarzwald als Zellstofffabrik für Strohzellstoff gegründet. Über die Spezialisierung auf Seiden-, Zigaretten- (1904) und Kondensatorpapiere (1928) steigt das Unternehmen zum Weltmarktführer für Overlay- (1955), Teebeutel- (1973) und Kaffeepadpapieren auf. Seine hochdiversifizierte Palette transparenter Overlays, die das Dekor unsichtbar schützen und der Oberfläche Abriebfestigkeit verleihen, hat wesentlichen Anteil am Siegeszug der Laminatoberfläche in den fünfziger und des Laminatfußbodens seit Beginn der neunziger Jahre. 1998 übernimmt der amerikanische Papierspezialist P. H. Glatfelter das Unternehmen und stärkt erneut dessen Innovationskraft.

Schoeller and Hoesch was founded as a cellulose factory for straw cellulose in Gernsbach in the Black Forest in 1881. Through the specialisation in silk, cigarette (1904) and condenser papers (1928) the company arose to become a world market leader for overlay papers (1955), tea bags (1973) and coffee pad papers. Its highly diversified range of transparent overlays, which invisibly protect the decor and give abrasion resistance to the surface, achieved an important share in the triumph of the laminate surface in the fifties and the laminate flooring since the beginning of the nineties. In 1998 the American paper specialist P. H. Glatfelter acquired the company and once again strengthened its innovative forces.

TOUGH IN RESISTANCE: TRANSPARENT SURFACE PROTECTION

SCHOELLER & HOESCH MANUFACTURE OVERLAY PAPERS FOR VARIOUS FURNITURE AND LAMINATE COMPOUND MATERIALS:

- Highly filled corundum-holding overlay papers for the manufacture of laminate flooring for all abrasion categories from AC 1 to AC 5, as well as highly abrasion-resistant coatings for OSB (Oriented Strength Boards) and plywood for the automotive and construction industries (substance from 18 to 75 g/m^2).
- Unfilled overlay papers for the manufacture of liquid-overlays for all manufacturing processes for LPL and HPL applications (substance from 18 to 35 g/m^2).
- Unfilled overlay papers for the manufacture of overlay melamine films for exterior applications (substance from 16 to 30 g/m^2).
- Filled overlay papers with a corundum content of < 10 % for the manufacture of HPL, CPL and LPL for worktops and laminated timber worktops with high exposure to abrasion.
- Filled and unfilled overlay papers for the manufacture of moulded timber (e. g. trays or moulded timber chairs).
- Natural coloured overlay papers for veneers and dark printed decors.

■ **ANMERKUNGEN**

Lattermann, S. 10–19

1 Resopal Forum, *Von der Schallplatte zur Schichtstoffplatte 100 Jahre H. Römmler GmbH.* 1962, S. 2
2 Wikipedia: *Schallplatte*. http://de.wikipedia.org/wiki/Schallplatte, 10.08.2006
3 *125 Jahre Forbo Resopal – Das Buch zum Jubiläum*, Resopal GmbH, Groß-Umstadt 1992, S. 133
4 *Römmlerbuch. Handbuch der Römmler Presstoffe*, H. Römmler AG, Spremberg 1938, S. 14
5 Brandenburger, Kurt: *Kunststoff Ratgeber*. Essen 1939, S. 36
6 Firmenschrift: *Fabrikationsprogramm der H. Römmler AG, Spremberg*, 1934
7 Firmenschrift: Griesshaber, *Erfahrungsbericht über die H. Römmler Aktiengesellschaft, Spremberg*, 1937
8 Firmenbericht *H. Römmler AG, Presstoffwerke*. Spremberg, 1936
9 *Römmlerbuch*, a.a.O., S. 94
10 Chronik der Fa. H. Römmler AG, 1939/40 (aus dem Besitz der ehemaligen Resopal-Mitarbeiterin Ruth Schmidt, Berlin; übergeben der Fa. Resopal, Groß-Umstadt, 1951), S. 14
11 *Römmlerbuch*, a.a.O., S. 94
12 Lattermann, Günter: »Bauhaus ohne Kunststoffe? – Kunststoffe ohne Bauhaus?« In: *form+zweck* 20, 2003, S. 110–127
13 Tilson, Barbara: *The development of the british plastics industry 1855 to 1990*, Centre for Urban and Regional Studies, Birmingham, 1999, S. 50
14 Brandenburger, Kurt: »Die Entwicklung der Aminoplastpresserei in Deutschland.« In: *Kunststoffe* 23, 1933, S. 5
15 Chronik, a.a.O., S. 12
16 Kohnert, E.: »Tischgerät aus Kunstharz-Presstoff.« In: *Die Schaulade* 6, 1930, S. 521–522
17 Werbeanzeige. In: *Die Schaulade*, 6, 1930, S. 734
18 Werbeanzeige. In: *Die Schaulade*, 6, 1930, S. 1067
19 Wiesenthal, Heinrich: »Chemisch hergestellte Kunststoffe auf der Leipziger Frühjahrsmesse 1931.« In: *Kunststoffe*, 21, 1931, S. 111
20 Aug. Hedinger GmbH & Co. http://www.hedinger.de/bilder/17/alresathed.pdf, 10.08.2006
21 Aug. Hedinger GmbH & Co. http://www.hedinger.de/chemtechpro/d_aktuelles. asp, 10.08.2006
22 Lutz, Peter: *Handelsnamen der Kunststoffe*. http://www.peterlutz.ch/lernen/werkstoff/kunststoffe/mkun10b.html, 10.08.2006
23 *Römmlerbuch*, a.a.O.
24 *125 Jahre*, a.a.O., S. 84
25 *125 Jahre*, a.a.O., S. 32, 35
26 Werbeanzeige. In: *Die Schaulade*, 7, 1931, S. 9
27 Werbeanzeige. In: Die Porzellan- und Glashandlung/Haus und Küchenmagazin 8, 1930, S. 297
28 Bossman, Andreas/Thöner, Wolfgang: »Das Bauhaus 1919 bis 1933 – Ursprünge und Vorgeschichte des Bauhauses.« In: Leismann, Burkhard (Hg): *Das Bauhaus*, Köln 1993, S. 16
29 Gropius, Walter: *Die Tragfähigkeit der Bauhausidee*. In: Wingler, Hans M.: *Das Bauhaus 1919–1933 Weimar Dessau Berlin*, Köln 1975, S. 120
30 Brandenburger, Kurt: *Herstellung und Verarbeitung von Kunstharzpressmassen*, Band 4, *Pressmischungen, geschichtete Produkte und deren Aufbereitung*, München/Berlin 1937, S. 11
31 Weber, Klaus (Hg): *Die Metallwerkstatt am Bauhaus*. Berlin 1992, S. 316
32 Kohnert, E.: *Tischgeräte aus Kunstharz-Presstoff*. In: *Die Schaulade* 6, 1930, S. 522
33 Hofmann, Beate Alice: *christian dell silberschmied und leuchtengestalter im 20. jahrhundert*. Katalog zur Ausstellung im Museum Hanau, Hanau 1996, S. 73
34 *Resopal Kerit*. Verkaufskatalog, Plastica GmbH, Berlin 1935
35 Wiesenthal, Heinrich: »Presstoffe und Verwandtes auf der Herbst-‚Messe' 1932.« In: *Kunststoffe* 22, 1932, S. 245
36 Hofmann, Beate, a.a.O., S. 73–74
37 Werbeanzeige. In: *Die Schaulade*, 10, 1934, S. 35
38 Lattermann, Günter: »Das Stapelservice von Christian Dell.« In: Buchholz, Kai/Wolbert, Klaus (Hg.): *Im Designerpark/Leben in künstlichen Welten*. Katalog zur Ausstellung im Institut Mathildenhöhe Darmstadt, Darmstadt 2004, S. 808–811
39 Lattermann, Günter: »Das Stapelservice von Christian Dell.« In: Breuer, Gerda (Hg.), *Designgeschichte ausstellen – Die Designsammlung der Universität Wuppertal*, Wuppertal 2005, S. 102–108
40 Vgl.: Neudecker, Irene: *75 Jahre Resopal – Ein Material schreibt Kulturgeschichte – 1930–2005*, FH Coburg 2005
41 Nachlass Jupp Ernst, Germanisches Nationalmuseum Nürnberg
42 Klein, Fr.: »Die Bedeutung der weißen Pressmasse ‚Resopal' für die Gestaltungsmöglichkeit moderner Schaltapparate und Beleuchtungskörper.« In: *Kunststoffe* 22, 1932, S. 78
43 Werbeprospekt, *Isolierstoff-Leuchten*, Stotz-Kontakt GmbH, Mannheim 1930/31
44 Sommerfeld, Arthur: *Plastische Massen*, Berlin 1934, S. 237
45 Vertrag zwischen D.A.G. und Römmler A.G. vom 01.07.1933
46 Werbeanzeige. In: *Die Schaulade*, 15, Ausgabe B 1939, S. 208
47 Pollopas + Reklamazia, Brandenburgisches Landeshauptarchiv Potsdam, Bestand/SignaturRep. 75, Chem. Werke H. Römmler AG, Nr. 18
48 Chronik, a.a.O., S. 16
49 Benn, Fritz: *Das Ende der H. Römmler AG in Spremberg*, Bericht des ehemaligen stellvertretenden Vorstandmitglieds der H. Römmler AG, Groß-Umstadt ca. 1954
50 Lattermann, Günter: *Bauhaus ohne Kunststoffe …*, a.a.O., S. 121
51 *VEB Sprela-Werke Spremberg*, Festbroschüre zum 25-jährigen Bestehen, Spremberg 1974, S. 27, 33
52 *VEB Sprela-Werke Spremberg*, a.a.O., S. 33
53 *125 Jahre Forbo Resopal*, a.a.O., S. 135
54 Ohlhauser, Gerd: »Grußwort der Markendesigner Jupp Ernst.« In: Breuer, Gerda (Hg.): *Jupp Ernst 1905–1987*, Bergische Universität Wuppertal 2006, S. 9–11

■ **NOTES**

Lattermann, pp. 10–19

1 Resopal Forum, *Von der Schallplatte zur Schichtstoffplatte 100 Jahre H. Römmler GmbH*, 1962, p. 2.
2 URL: http://de.wikipedia.org/wiki/Schallplatte
3 *125 Jahre Forbo Resopal – Das Buch zum Jubiläum*, Resopal GmbH, Groß-Umstadt 1992, p. 133.
4 *Römmlerbuch. Handbuch der Römmler Presstoffe*, H. Römmler AG, Spremberg N.-L. 1938, p. 14.
5 Brandenburger, Kurt, *Kunststoff Ratgeber*, Essen 1939, p. 36.
6 Company publication: *Fabrikationsprogramm der H. Römmler A.-G. Spremberg ND.-L.*, 1934.
7 Company publication: Griesshaber, *Erfahrungsbericht über die H. Römmler Aktiengesellschaft, Spremberg*, 1937.
8 Firmenbericht *H. Römmler A.G., Presstoffwerke*, Spremberg 1936.
9 *Römmlerbuch*, op. cit., p. 94.
10 Chronik der Fa. H. Römmler AG, 1939/40 (owned by former Resopal employee, Ruth Schmidt, Berlin; donated to Resopal, Groß-Umstadt, 1951), p. 14.
11 *Römmlerbuch*, op. cit., p. 94.
12 Lattermann, Günter, "Bauhaus ohne Kunststoffe? – Kunststoffe ohne Bauhaus?", in *form+zweck* 20, (2003), p. 110–127.
13 Tilson, Barbara, *The development of the British plastics industry 1855 to 1990*, Centre for Urban and Regional Studies, Birmingham, 1999, p. 50.
14 Brandenburger, Kurt, "Die Entwicklung der Aminoplastpresserei in Deutschland", in *Kunststoffe* 23, (1933), p. 5.
15 Chronik, op. cit., p. 12.
16 Kohnert, E., "Tischgerät aus Kunstharz-Presstoff" in *Die Schaulade* 6, (1930), pp. 521-522.
17 Advertisement; in *Die Schaulade*, 6, (1930), p. 734.
18 Advertisement; in *Die Schaulade*, 6, (1930), p. 1067.
19 Wiesenthal, Heinrich, "Chemisch hergestellte Kunststoffe auf der Leipziger Frühjahrsmesse 1931", in *Kunststoffe*, 21 (1931), pp. 110–111.
20 http://www.hedinger.de/bilder/17/alresathed.pdf
21 URL: http://www.hedinger.de/chemtechpro/d_aktuelles.asp
22 URL: http://www.peterlutz.ch/lernen/werkstoff/kunststoffe/mkun10b.html
23 *Römmlerbuch*, op. cit.
24 *125 Jahre*, op. cit., p. 84.
25 *125 Jahre*, op. cit., pp. 32,35.
26 Advertisement; in *Die Schaulade*, 7, (1931), p. 9.
27 Advertisement, in *Die Porzellan- und Glashandlung/Haus und* Küchenmagazin, 8, (1930), p. 297.
28 Bossman, Andreas/Thöner, Wolfgang, "Das Bauhaus 1919 bis 1933 – Ursprünge und Vorgeschichte des Bauhauses", in: Leismann, Burkhard (ed.), *Das Bauhaus*, Cologne 1993, p. 16.
29 Gropius, Walter, "Die Tragfähigkeit der Bauhausidee", in: Wingler, Hans M., *Das Bauhaus 1919–1933 Weimar Dessau Berlin*, Cologne 1975, p. 120.
30 Brandenburger, Kurt, *Herstellung und Verarbeitung von Kunstharzpressmassen*, Vol. 4 *Pressmischungen, geschichtete Produkte und deren Aufbereitung*, Munich/Berlin 1937, p. 11.
31 Weber, Klaus (ed.), *Die Metallwerkstatt am Bauhaus*, Berlin 1992, p. 316.
32 Kohnert, E. *Tischgeräte aus Kunstharz-Presstoff*; in *Die Schaulade* 6, (1930), p. 522.
33 Hofmann, Beate Alice, *christian dell silberschmied und leuchtengestalter im 20. jahrhundert*, catalogue for an exhibition in Museum Hanau, Hanau 1996, p. 73.
34 *Resopal Kerit*. sales catalogue, Plastica GmbH, Berlin 1935.
35 Wiesenthal, Heinrich, "Presstoffe und Verwandtes auf der Herbst-‚Messe' 1932", in *Kunststoffe* 22, (1932), p. 245.
36 Hofmann, Beate Alice, op. cit., pp. 73–74.
37 Advertisement; in *Die Schaulade*, 10, (1934), p. 35.
38 Lattermann, Günter, "Das Stapelservice von Christian Dell", in Buchholz, Kai/Wolbert, Klaus (eds.), *Im Designerpark/Leben in künstlichen Welten*, catalogue for an exhibition at the Institut Mathildenhöhe Darmstadt, Darmstadt 2004, pp. 808–811.
39 Lattermann, Günter, "Das Stapelservice von Christian Dell", in Breuer, Gerda (ed.), *Designgeschichte ausstellen – Die Designsammlung der Universität Wuppertal*, Wuppertal 2005, pp. 102–108.
40 Cf..: Neudecker, Irene, *75 Jahre Resopal – Ein Material schreibt Kulturgeschichte – 1930–2005*, FH Coburg 2005.
41 Legacy Jupp Ernst, Germanisches Nationalmuseum Nürnberg.
42 Klein, Fr., „Die Bedeutung der weißen Pressmasse ‚Resopal' für die Gestaltungsmöglichkeit moderner Schaltapparate und Beleuchtungskörper", in *Kunststoffe* 22, (1932), p. 78.
43 Werbeprospekt, *Isolierstoff-Leuchten*, Stotz-Kontakt GmbH, Mannheim 1930/31.
44 Sommerfeld, Arthur, *Plastische Massen*, Berlin 1934, p. 237.
45 Contract between D.A.G. and Römmler A.G. of 1. 7. 1933.
46 Advertisement; in *Die Schaulade*, 15, Ausgabe B (1939), p. 208.
47 Pollopas + Reklamazia, Brandenburgisches Landeshauptarchiv Potsdam, Bestand/SignaturRep. 75, Chem. Werke H. Römmler AG, No. 18.
48 Chronik, op. cit., p. 16.
49 Benn, Fritz, *Das Ende der H. Römmler AG in Spremberg*, Report of former deputy member of the board of H. Römmler AG, Groß-Umstadt ca. 1954.
50 Lattermann, Günter: Bauhaus ohne Kunststoffe …, op. cit., p. 121.
51 *VEB Sprela-Werke Spremberg*, brochure commemorating the 25th anniversary of the company, Spremberg 1974, pp. 27, 33.
52 *VEB Sprela-Werke Spremberg*, op. cit., p. 33.
53 *125 Jahre Forbo Resopal*, op. cit., p. 135.
54 Ohlhauser, Gerd, "Grußwort – Der Markendesigner Jupp Ernst", in Breuer, Gerda (ed.), *Jupp Ernst 1905–1987*, Bergische Universität Wuppertal 2006, pp. 9–11.
55 *125 Jahre Forbo Resopal*, op. cit., p. 136.

55 *125 Jahre Forbo Resopal*, a.a.O., S. 136
56 Hofmann, Beate Alice, a.a.O., S. 15
57 Hofmann, Beate Alice, a.a.O., S. 17
58 Scheiffele, Walter: »Der rote Koffer – ein Ulmer Modell.« In: Petruschat, Angelika (Hg.): *macht langsam … – 50 Jahre Peter Raacke Design*, form+zweck Verlag, Berlin 2004, S. 36
59 Pfaender, Heinz G.: *Meine Zeit in der Werkstatt Wagenfeld, Tagebuch 1954–1957*, Hamburg 1998, S. 15
60 Hofmann, Beate Alice, a.a.O., S. 23–24, 72–74
61 Lattermann, Günter: *Bauhaus ohne Kunststoffe* …, a.a.O., S. 122–123
62 Leonhardt, Brigitte: »Objekte aus Kunststoff von Friedrich Adler.« In: Leonhardt, Brigitte/Götz, Norbert/Zühlsdorf, Dieter (Hg.): *Friedrich Adler – zwischen Jugendstil und Art Deco*, Stuttgart 1994, S. 384–391
63 Lattermann, Günter: »Der Volksempfänger.« In: Buchholz, Kai/Wolbert, Klaus (Hg.): *Im Designerpark/Leben in künstlichen Welten*, Katalog zur Ausstellung im Institut Mathildenhöhe Darmstadt, Darmstadt 2004, S. 940–941
64 Cremer-Thursby, Marc: *Design der dreißiger und vierziger Jahre in Deutschland – Hermann Gretsch*, Frankfurt am Main 1996, S. 185–186, 295–296

Breuer, S. 38–51
1 Bartning, Otto: »Kitsch ist Lebensangst.« In: *Die Neue Zeit*. Nr. 162, 11. Juli 1950
2 Popp, Josef: »Einführung.« In: Dürerbund-Werkbund-Genossenschaft (Hrsg.): *Deutsches Warenbuch*. Hellerau bei Dresden o.J. (1915), S. XXXII
3 Christian Dells Plastikgeschirr aus Resopal oder László Moholy-Nagys Vorliebe für Plexiglas in seinen Kunstwerken sollten nicht darüber hinwegtäuschen, dass es sich um Ausnahmen handelte.
4 Der Prototyp ging zwar erst 1990 in Produktion, wurde aber durch die genannte Ausstellung sehr bekannt.
5 Schwippert, Hans: *Rede auf dem Darmstädter Gespräch »Mensch und Technik«* 1952. Zit. nach ders.: »Das Ende der Werkgerechtigkeit.« In: *Baukunst und Werkform*, Heft 5, 1953, S. 236
6 Sedlmayr, Hans: *Verlust der Mitte*. Zit. nach Oestreich, Hans Dieter: »Abstieg zum Anorganischen.« In: *Baukunst und Werkform*, Heft 5, 1953, S. 237
7 Über Vorläufer und Verwendung der synthetischen Stoffe im 19. Jahrhundert, vgl. Ulmer, Renate: »Vom Historismus zur Moderne. Die ersten Kunststoffe und ihre Verwendung im 19. Jahrhundert.« In: Ausstellungskatalog Neue Sammlung, München und Museum Künstlerkolonie, Mathildenhöhe Darmstadt, 1997, S. 8–19
8 Semper, Gottfried: *Der Stil in den technischen und tektonischen Künsten*. 2 Bde., hier Bd. 1, München 1863, S. 44
9 Ebd.
10 Ruskin, John: »Die sieben Leuchter der Baukunst.« In: Kemp, Wolfgang, Dortmund 1994, S. 325
11 Ruskin, John, a.a.O., S. 316
12 »…es ist nicht das Material, sondern der Mangel an menschlicher Arbeit, der ein Ding wertlos macht. Ein Stück Terracotta oder Stuck, mit der Hand bearbeitet, wiegt alle Steine von Carrara, die mit der Maschine geschnitten sind, auf.« In: Ruskin, John, a.a.O., S. 317
13 Ruskin, John, a.a.O., S. 102
14 Ab 1860 wandte sich Dresser, der in absentia in Jena den Doktortitel für Botanik aufgrund seiner Buchveröffentlichungen und einer dort eingerechten Abhandlung über Pflanzenmorphologie erhielt, der praktischen Anwendung seines botanischen Grundwissens auf die künstlerische Gestaltung zu. Von den morphologischen Gesetzen der Pflanze leitete er die Geometrisierung der vegetabilen Form als oberste Prämisse des Designs ab – diese finde qua Ordnung (»order«) im Ornament ihre Anwendung. »Order« wurde somit zum Beweis für die Teilhaftigkeit menschlichen Intellekts am künstlerischen Gestaltungsprozess.
15 Dresser, Christopher: »Principles.« In: *Christopher Dresser. Ein viktorianischer Designer 1834–1904*. Katalog des Kunstgewerbemuseum der Stadt Köln, Köln 1981, S. 16
16 Vgl. über die Vertreter der »Guten Form« in den fünfziger und sechziger Jahren: Erni, Peter: *Die gute Form. Eine Aktion des Schweizerischen Werkbundes. Dokumentation und Interpretation*. Baden 1983, bes. das Kapitel *Die sündigen Gewänder des profanen Geschmacks: Ethik*
17 Brief von Gustav B. von Hartmann an Prof. Walter Rossow, Deutscher Werkbund e.V., Berlin. Werkbund Archiv Berlin, ADO 7 – 439/59a
18 Schwippert, Hans: »Das Ende der Werkgerechtigkeit.« In: *Baukunst und Werkform*, Heft 5, 1953, S. 235–236
19 Aus einer Rede Schwipperts auf dem Darmstädter Gespräch »Mensch und Technik« 1952, zitiert nach Schwippert, Hans, a.a.O.
20 Ebd.
21 Gustav B. von Hartmann anlässlich der Sitzung des Arbeitskreises für industrielle Formgebung am 4. August 1955 in Selb/Bayern, Werkbund Archiv Berlin, Akte »G. B. von Hartmann«
22 Schwabe, Amtor: »Was fängt der Architekt mit den Kunststoffen an?« In: *Baukunst und Werkform*, Heft 5, 1953, S. 244
23 Braun-Feldweg, Wilhelm: *Normen und Formen industrieller Produktion*. Ravensburg 1954, S. 135
24 Braun-Feldweg, Wilhelm, a.a.O., S. 145
25 Oestreich, Hans Dieter: »Abstieg zum Anorganischen?« In: *Baukunst und Werkform*, Heft 5, 1953, S. 240
26 Ebd.
27 Ernst, Jupp: »Kunststoffe sind unser Schicksal oder: Natur und Unnatur der Stoffe.« In: *Süddeutsche Zeitung* Nr. 101. Mittwoch, 28. April 1971, zit. nach Ernst, Jupp: *Designprobleme*. Faksimiles aus der Presse 1947–1975, Eigendruck ohne Jahr
28 Zusammenfassender Bericht über das Referat *Die Oberflächengestaltung von Kunststoffen* von Gustav B. von Hartmann anlässlich der Sitzung des Arbeitskreises für industrielle Formgebung am 4. August 1955 in Selb/Bayern, S. 1–2
29 Schwippert, Hans, a.a.O., S. 236

56 Hofmann, Beate Alice, op. cit., p. 15.
57 Hofmann, Beate Alice, op. cit., p. 17.
58 Scheiffele, Walter, "Der rote Koffer – ein Ulmer Modell" in Petruschat, Angelika (ed..), *"macht langsam …" – 50 Jahre Peter Raacke Design*, form+zweck Verlag, Berlin 2004, p 36.
59 Pfaender, Heinz G., *Meine Zeit in der Werkstatt Wagenfeld, Tagebuch 1954–1957*, Hamburg 1998, p. 15.
60 Hofmann, Beate Alice, op. cit., pp. 23–24, 72–74.
61 Lattermann, Günter, "Bauhaus ohne Kunststoffe …, op. cit., pp. 122–123.
62 Leonhardt, Brigitte, "Objekte aus Kunststoff von Friedrich Adler", in Leonhardt, Brigitte/Götz, Norbert/ Zühlsdorf, Dieter (eds.) *Friedrich Adler – zwischen Jugendstil und Art Deco*, Stuttgart 1994, pp. 384–391.
63 Lattermann, Günter, "Der Volksempfänger", in Buchholz, Kai/Wolbert, Klaus (eds.), *Im Designerpark/Leben in künstlichen Welten*, catalogue for an exhibition at the Institut Mathildenhöhe Darmstadt, Darmstadt 2004, pp. 940–941.
64 Cremer-Thursby, Marc, *Design der dreißiger und vierziger Jahre in Deutschland – Hermann Gretsch*, Frankfurt am Main 1996, pp. 185–186, 295–296.

Breuer, pp. 38–51
1 Bartning, Otto: "Kitsch ist Lebensangst", in: Die Neue Zeit, No. 162, 11 July 1950.
2 Popp, Josef: "Introduction", in: Dürerbund-Werkbund-Genossenschaft (ed.): Deutsches Warenbuch, Hellerau bei Dresden, year not given (1915), p. XXXII.
3 Christian Dell's plastic tableware made of Resopal or László Moholy-Nagy's love of Plexiglas in his works of art could give the wrong impression: they were, in fact, exceptions.
4 Although the prototype did not go into production until 1990, the exhibition made it very famous.
5 Schwippert, Hans: Speech given at the Darmstädter Gespräch "Man and Technology" in 1952, quoted in Hans Schwippert: "Das Ende der Werkgerechtigkeit", in: Baukunst und Werkform, no. 5, 1953, Sp. 236.
6 Sedlmayr, Hans: Verlust der Mitte, quoted in Hans Dieter Oestreich: "Abstieg zum Anorganischen?" in: Baukunst und Werkform, no. 5, 1953, p. 237.
7 On precursors to synthetic materials and their use in the 19th century cf. Renate Ulmer: "Vom Historismus zur Moderne. Die ersten Kunststoffe und ihre Verwendung im 19. Jahrhundert", in: catalogue of the Neue Sammlung, Munich and Museum Künstlerkolonie, Mathildenhöhe Darmstadt, 1997, pp. 8–19.
8 Semper, Gottfried: Der Stil in den technischen und tektonischen Künsten, 2 vols., here vol. 1, Munich 1863, p. 44.
9 Ibid.
10 Ruskin, John: The Seven Lamps of Architecture, New York 1989, p. 173 (Ger 325). A republication of the second edition of the work as published by George Allen, Sunnyside, Orpington, Kent, in 1880.
11 Ibid, p. 169 (Ger 316).
12 "For it is not the material, but the absence of the human labour, which makes the thing worthless; and a piece of terra cotta, or of plaster of Paris, which has been wrought by the human hand, is worth all the stone in Carrara, cut by machinery." pp. 55/56.
13 Ibid, Aphorism 15, in "Seven Lamps …", p. 56.
14 From 1860, Dresser, who had been awarded a doctorate in botany *in absentia* from Jena on the basis of books published and a treatise he submitted to the university on plant morphology, turned his attention to the possibility of applying his botanical knowledge to artistic design. He derived the geometrical stylisation of vegetable forms from the morphological laws of plants, making it the highest design premise – it is then used to give "order" to ornamentation. "Order" thus became proof of the participation of the human intellect in the artistic design process.
15 Dresser, Christopher: Principles of Victorian Decorative Design, New York, 1995, p. 16. A republication of Principles of Decorative Design, second edition, London, n.d.
16 Cf.. on the advocates of "good design" in the 1950s and 1960s: Peter Erni: Die gute Form. Eine Aktion des Schweizerischen Werkbundes. Dokumentation und Interpretation. Baden 1983, in particular the chapter "Die sündigen Gewänder des profanen Geschmacks: Ethik."
17 Letter from Hartmann to Prof. Walter Rossow, Deutscher Werkbund e.V., Berlin. Werkbund Archives Berlin, ADO 7 – 439/59a.
18 Schwippert, Hans: "Das Ende der Werkgerechtigkeit", in: baukunst und werkform, no. 5, 1953, p. 235–236.
19 From a speech by Schwippert at the Darmstädter Gespräch "Man and Technology" 1952, quoted in Hans Schwippert: Das Ende der Werkgerechtigkeit, op. cit.
20 Ibid.
21 Gustav B. von Hartmann at the meeting of the Arbeitskreis für industrielle Formgebung (Working Group on Industrial Design) on 4 August 1955 in Selb/Bavaria, Werkbund archives Berlin.
22 Schwabe, Amtor: "Was fängt der Architekt mit den Kunststoffen an?" in: Baukunst und Werkform, no. 5, 1953, pp. 242–250, here p. 244.
23 Braun-Feldweg, Wilhelm: Normen und Formen industrieller Produktion, Ravensburg 1954, p. 135.
24 Braun-Feldweg: op. cit., p. 145.
25 Oestreich, Hans Dieter: "Abstieg zum Anorganischen?" in: Baukunst und Werkform, no. 5, 1953, pp. 237–241, here p. 240.
26 Ibid.
27 Ernst, Jupp: "Kunststoffe sind unser Schicksal oder: Natur und Unnatur der Stoffe", in: Süddeutsche Zeitung No. 101. Wednesday, 28 April 1971, quoted from Jupp Ernst: Designprobleme. Facsimiles from the press 1947–1975, self-published, no year given.
28 Summary of the paper "Oberflächengestaltung von Kunststoffen" given by G.B. von Hartmann at the meeting of the Arbeitskreis für industrielle Formgebung (Working Group on Industrial Design) on 4 August 1955 in Selb/Bavaria, pp. 1–2.
29 Schwippert, Hans: Speech at the Darmstädter Gespräch op. cit., p. 236.

Brachert, S. 92–101

1 Deutscher Sparkassenverlag (Hg.): *Im Reich der Frau*. In: *Wie bauen – wie wohnen*. Heft 5, Stuttgart 1956, S. 4–5
2 Brachert, Eva: *Hausrat aus Plastic. Alltagsgegenstände aus Kunststoff in Deutschland in der Zeit von 1950–1959*. Weimar 2002, S. 160 ff.
3 Werbeslogan der Leicht-Küchenwerbung. In: *Die Kunst und das schöne Heim*. 1953, S. 178
4 »Hat Monika nicht recht?« In: *Kunststoff-Berater*. Februar 1957, S. 63
5 Brachert, Eva, a.a.O., S. 176
6 Vgl.: Brödner, Erika: *Moderne Küchen*. München 1950
7 Brachert, Eva: *Hausrat aus Plastic*. Weimar 2002, S. 29 ff.
8 Innovatives Design, wie die von Christian Dell 1933/35 gestalteten Waren aus Resopal, wurden nicht erwähnt. Da Messeberichte und Prospektmaterial nicht als Indikator für die tatsächliche Verbreitung dieser Produkte gelten können, muss auch hier die Frage gestellt werden: Wer besaß diese gestalterisch innovativen Kunststoffartikel? Vgl.: Gretsch, Hermann: *Hausrat der zu uns passt*. Stuttgart 1940; Brachert, Eva: *Hausrat aus Plastic*. Weimar 2002, S. 44ff.
9 Vgl.: *Der Plastverarbeiter*. 1955, S. 75
10 Margarete Schütte-Lihotzky beklagte diesen Fakt bereits 1927. Vgl: Noever, Peter (Hg.): *Die Frankfurter Küche von Margarete Schütte-Lihotzky*. Wien 1997
11 Dafür war das »Amt für Schönheit der Arbeit« (gegr. 1933) von Staats wegen zuständig, das ab 1935 die »Deutsche Warenkunde« herausgab.
12 Vgl.: Manske, Beate (Hg.): *Wie Wohnen – Von Lust und Qual der richtigen Wahl. Ästhetische Bildung in der Alltagskultur des 20. Jahrhunderts*. Bremen 2004
13 Es gab deutschlandweit 75 autorisierte WK-Einrichtungshäuser, die alle über die gleichen WKS-Markenmöbel verfügten.
14 Neben Jupp Ernst waren Wera Meyer-Waldeck, Ernst Oberhoff und Helmut Lortz für den Messeauftritt verantwortlich.
15 *Constanze – Schöner wohnen – Schöner bauen*. Nr. 27, IV 1956
16 Schwabe, A.: »Warum Kunststoffe.« In: Haus und Heim, Sonderheft: Kunststoffe und Plasticgeräte für den Haushalt, Hamburg 1958, S. 5

Höhns, S. 144–155

1 Vgl. Semper, Gottfried: *Der Stil in den technischen und tektonischen Künsten oder praktische Aesthetik (1860–1863)*. Mittenwald 1977
2 Loos, Adolf: *das prinzip der bekleidung*. In: Ders.: *Ins Leere gesprochen*. Wien/München 1962, S. 106f.
3 Moos, Stanislaus von: »Glanzblitz! Über Architektur, Transparenz und Multimedialität.« In: Wagner, Monika/Rübel, Dietmar (Hg.): *Material in Kunst und Alltag*. Berlin 2002, S. 46. Von Moos zitiert dabei selbst einen Brief des in Barcelona bauleitenden Architekten Kettler an Lilly Reich vom 10.09.1929, der sich im Mies van der Rohe-Archiv im Museum of Modern Art in New York befindet.
4 Höhns, Ulrich: »L´apparance ne trompe pas. Robert Mallet-Stevens, L´ouvre complète.« In: *werk, bauen + wohnen*, Heft 10, Zürich 2005, S. 56f.
5 Vgl. Fuhrmeister, Christian: »Ewige Materie in dünnen Platten. Naturstein in der zeitgenössischen Architektur.« In: *Material in Kunst und Alltag*, a.a.O.
6 Vgl. »Ehemalige Taschentuchweberei in Blumberg.« In: Grimm, Friedrich/Richarz, Clemens: *Hinterlüftete Fassaden. Konstruktionen vorgehängter hinterlüfteter Fassaden aus Faserzement*. Stuttgart/Zürich 1994
7 Einladung zur Vortragsveranstaltung *Faszination Bauen* zur geplanten Elbphilharmonie am 22.03.2006, Technische Universität Hamburg-Harburg und ReGe Hamburg Projekt-Realisierungsgesellschaft mbH
8 Pehnt, Wolfgang: *Deutsche Architektur seit 1900*. München 2006, S. 488
9 Vgl. Colquhoun, Alan: »Die Fassade in ihren modernen Varianten.« In: *werk, bauen + wohnen*, Heft 12, Zürich 2005
10 Vgl. Hoffmann, Kurt/Griese, Helga/Meyer-Bohe, Walter: *Fassaden. Form und Detail von Außenwänden und Außenwandbekleidungen*. Stuttgart 1973
11 Vgl. Hindrichs, Dirk U./Heusler, Winfried (Hg.): *Fassaden – Gebäudehüllen für das 21. Jahrhundert* (Schüco International KG), Basel 2004
12 Wolfgang Bachmann im Gespräch mit Markus Allmann. In: *Über Oberflächen*. Baumeister Heft 3, München 2004, S. 55

Selle, S. 168–179

1 Vgl. Heubach, Friedrich W.: *Das bedingte Leben. Theorie der psychologischen Gegenständlichkeit der Dinge. Ein Beitrag zur Psychologie des Alltags*. München 1996, S. 129f.
2 Resopal GmbH (Hg.): *Resopal 2008*. Groß-Umstadt 2003, S. 16
3 Ebd.
4 Resopal GmbH (Hg.): *QWEAR. Resopal 2005*. Groß-Umstadt 2005, S. 42
5 *QWEAR. Resopal 2005*, a.a.O., S. 6
6 Baudrillard, Jean: *Agonie des Realen*. Berlin 1978, S. 25
7 *Resopal 2008*, a.a.O., S. 140
8 *Resopal 2008*, a.a.O., S. 5
9 Fischer, Volker: *Der Traum aus dem die Stoffe sind. Das Geheimnis der Eigenschaften. »ent.materialien« von Tom Stark*. Frankfurt 2006, o. S.
10 *Resopal 2008*, a.a.O., S. 138
11 *QWEAR. Resopal 2005*, a.a.O., S. 3

Brachert, pp. 92–101

1 Deutscher Sparkassenverlag (eds.): *Im Reich der Frau*. In: *Wie bauen – wie wohnen*. no. 5, Stuttgart 1956, pp. 4–5
2 Brachert, Eva: *Hausrat aus Plastic*. Weimar 2002, p. 160 ff.
3 Advertising slogan for Leicht kitchens. In: *Die Kunst und das schöne Heim*. 1953, p. 178.
4 "Hat Monika nicht recht?" In: *Kunststoff-Berater*. February 1957, p. 63.
5 Brachert, Eva: *Hausrat aus Plastic. Alltagsgegenstände aus Kunststoff in Deutschland in der Zeit von 1950–1959*. Weimar 2002, p. 176.
6 Cf.: Brödner, Erika: *Moderne Küchen*. Munich 1950.
7 Brachert, Eva: *Hausrat aus Plastic*. Weimar 2002, p. 29 ff.
8 Innovative design, such as the goods Christian Dell designed in Resopal in 1933/35, were not mentioned. Since trade show reports and brochures cannot be taken as an indicator for how widespread these products actually were, the question must be asked: who owned these plastic articles with their innovative design? Cf.: Gretsch, Hermann: *Hausrat der zu uns passt*. Stuttgart 1940; Brachert, Eva: *Hausrat aus Plastic*. Weimar 2002, p. 44 ff.
9 Cf.: *Der Plastverarbeiter*. 1955, p. 75.
10 Margarete Schütte-Lihotzky had bemoaned this fact back in 1927. Cf.: Noever, Peter (ed.): *Die Frankfurter Küche von Margarete Schütte-Lihotzky*. Vienna 1997.
11 The "Amt für Schönheit der Arbeit" (Office for the Beauty of Work), which was established in 1933 was the state agency responsible. From 1935 it published "Deutsche Warenkunde."
12 Cf.: Manske, Beate (ed.): *Wie Wohnen – Von Lust und Qual der richtigen Wahl. Ästhetische Bildung in der Alltagskultur des 20. Jahrhunderts*. Bremen 2004.
13 There were 75 authorized WK furniture stores throughout Germany. They all had the same WKS brand furniture.
14 As well as Jupp Ernst, Wera Meyer-Waldeck, Ernst Oberhoff and Helmut Lortz were also responsible for the stand at the trade show.
15 *Constanze – Schöner wohnen – Schöner bauen*. No. 27, IV 1956.
16 Schwabe, A.: "Warum Kunststoffe." In: Haus und Heim, special issue: Kunststoffe und Plasticgeräte für den Haushalt, Hamburg 1958, p. 5.

Höhns, pp. 144–155

1 Cf. Semper, Gottfried: *Style in the technical and tectonic arts; or practical aesthetics* (1860–1863), 2 volumes, reprinted: Mittenwald 1977.
2 Loos, Adolf: "The Principle of Cladding", in: *Spoken Into the Void: Collected Essays 1897–1900*, Cambridge, Mass. 1982.
3 von Moos, Stanislaus: "'Glanzblitz!'. Über Architektur, Transparenz und Multimedialität", in: Monika Wagner and Dietmar Rübel (eds.), *Material in Kunst und Alltag*, Berlin 2002, p. 46. Von Moos quotes a letter from Kettler, the site manager in Barcelona to Lilly Reich, dated 10.9.1929, which is now in the Mies van der Rohe Archive at the Museum of Modern Art in New York.
4 Höhns, Ulrich: "L´apparance ne trompe pas." Robert Mallet-Stevens, L´œuvre complète, Édition du Centre Pompidou, Paris 2005, in: *werk, bauen + wohnen*, no. 10, Zurich 2005, p. 56 f.
5 Cf. Fuhrmeister, Christian: "'Ewige Materie' in dünnen Platten. Naturstein in der zeitgenössischen Architektur", in: *Material in Kunst und Alltag*, op. cit.
6 Cf. "Ehemalige Taschentuchweberei in Blumberg", in: Friedrich Grimm, Clemens Richarz: *Hinterlüftete Fassaden. Konstruktionen vorgehängter hinterlüfteter Fassaden aus Faserzement*, Stuttgart/Zurich 1994.
7 Invitation to a public lecture entitled "The fascination of building" on the planned Elbphilharmonie on 22.3.2006, Technische Universität Hamburg-Harburg and ReGe Hamburg Projekt-Realisierungsgesellschaft mbH.
8 Pehnt, Wolfgang: *Deutsche Architektur seit 1900*, section on "Faszination der Oberflächen", Munich, 2nd edition 2006, p. 488.
9 Cf. Colquhoun, Alan: "Die Fassade in ihren modernen Varianten", in: *werk, bauen + wohnen*, Zurich, no. 12/2005.
10 Cf. Hoffmann, Kurt; Griese, Helga; Meyer-Bohe, Walter: *Fassaden. Form und Detail von Außenwänden und Außenwandbekleidungen*, Stuttgart 1973.
11 Cf. Hindrichs, Dirk U.; Heusler, Winfried (eds.): *Fassaden – Gebäudehüllen für das 21. Jahrhundert* (Schüco International KG), Basel 2004.
12 Wolfgang Bachmann talking to Markus Allmann, in: "Über Oberflächen". *Baumeister*, no. 3, Munich 2004, p. 55.

Selle, pp. 168–179

1 Cf. Heubach, Friedrich W.: *Das bedingte Leben. Theorie der psychologischen Gegenständlichkeit der Dinge. Ein Beitrag zur Psychologie des Alltags*. Munich, 2nd edition 1996, pp. 129 f.
2 Resopal GmbH (ed.): *Resopal 2008*. Groß-Umstadt 2003, p. 16.
3 Ibid.
4 Resopal GmbH (ed.): *QWEAR. Resopal 2005*. Groß-Umstadt 2005, p. 42.
5 Ibid, p.6.
6 Baudrillard, Jean: *Agonie des Realen*. Berlin 1978, p. 25.
7 *Resopal 2008*, op. cit., p. 140.
8 *Resopal 2008*, op. cit., p. 5.
9 Fischer, Volker, in: *Der Traum aus dem die Stoffe sind. Das Geheimnis der Eigenschaften. "ent.materialien" by Tom Stark*. Catalogue. Museum für Angewandte Kunst Frankfurt, 2006, pages not numbered.
10 *Resopal 2008*, op. cit., p.138.
11 *QWEAR. Resopal 2005*, op. cit., p. 3.

■ **VERWENDETE QUELLEN BEI DEN KURZTEXTEN**
BIBLIOGRAPHY FOR THE CATALOGUE SECTIONS

Arbeitsgemeinschaft Hauswirtschaft e.V. (Hg.): *Haushaltsträume – Ein Jahrhundert Technisierung und Rationalisierung im Haushalt*, Königstein im Taunus 1990

Bergische Universität Wuppertal, Fachbereich F Architektur, Design, Kunst, Fachgebiet Kunst- und Designgeschichte: Korrespondenz von Jupp Ernst mit der H. Römmler GmbH von 1952-1967

Brachert, Eva: *„Hausrat aus Plastic" – Alltagsgegenstände aus Kunststoff in Deutschland in der Zeit von 1950-1959*, Weimar 2002

Brandenburgisches Landeshauptarchiv (BLHA), Potsdam: Rep. 75 Chem. Werk Römmler, 33 Akten, Zeitraum 1933-1945; Rep. 203 AzS BET 1144 (Enteignungsakte der Firma); Rep. 250 Landratsamt Spremberg, Nr. 276 (Überführung des VEB in das KWU Spremberg); Rep. 271 VVB Chemie und Papier Nr. 21 (Rechtsangelegenheiten), Nr. 283-286 (Bilanzen und Inventuren des VEB Pressstoffwerk Spremberg 1949-1950).

Breuer, Gerda (Hg.): *Designgeschichte ausstellen. Die Designsammlung der Universität Wuppertal*, Hagen 2005

Breuer, Gerda (Hg.): *Sarah Pelikan – Material Resopal. Arbeiten 1982-1997*, Ausstellungskatalog Bergische Universität Wuppertal 4.11.-9.12.2005, Wuppertal 2005

Buck, Alex: *Dominanz der Oberfläche. Betrachtungen zu einer neuen Bedeutsamkeit der Gegenstände*, Frankfurt am Main 1998

Detail: Material und Oberfläche, Nr. 6, 2006

Forbo-Resopal GmbH (Hg.): *Resopal-Handbuch. Anwendungs- und Verarbeitungsempfehlungen, Technische Hinweise, Tabellen und technische Daten*, Groß-Umstadt 1997, 4. Auflage

Griesdorn, Alfons/Leufer, Heinz: *Die sichtbaren Möbel-Flächen – unter besonderer Berücksichtigung der Küchenmöbel*, Darmstadt 1984

Grieshaber, HAP: *Seestern und Tomahawk. Vierzehn Collagen unter Resopal. Zwei Farbholzschnitte*, Insel-Bücherei Nr. 849, Frankfurt am Main 1965

ID4, für die Resopal GmbH, Groß-Umstadt (Hg.): *Die Sinnlichkeit der Oberfläche. Produkte, Materialien und Verfahren: Wie Oberflächen unsere Welt prägen*, Darmstadt 2006

Kunstverein Darmstadt e.V. (Hg.): *Magie einer alltäglichen Moderne. Historische Kunststoffobjekte der Sammlung Kölsch*, Ausstellungskatalog Kunsthalle Darmstadt 24.8-21.9.1986, Heidelberg 1986

Neues Museum in Nürnberg (Hg.): *Richard Artschwager. Up and Across*, Nürnberg 2001

Polster, Bernd/Meyer, Olaf: *Braun – 50 Jahre Produktinnovationen*, Köln 2005

H. Römmler AG (Hg.): *Römmlerbuch*, Düsseldorf 1938

H. Römmler GmbH (Hg.): *Resopal Forum. Blätter für eine freie Diskussion*, Groß-Umstadt, o. J.
Mut zum Dekor, [1961]
Nr. 4: o. T., [1962]
Nr. 5: o. T., [1963]
Nr. 6: *Neue Wege der Dekoration. Resopal Unterdruck*, [1964]
Nr. 7: *Das Büro-Zeitalter*, [1964]
Nr. 12 *Pro Juventute*, [1964]
Nr. 13: *Resopal auf allen Wegen*, [1964]
Nr. 14: *Erfahrung mit Resopal*, [1964]
Nr. 15: *Resopal für die Kunst*, [1966]
Von der Schallplatte zur Schichtstoffplatte. 100 Jahre H. Römmler GmbH, [1967]
Nr. 17: *Resopal im Hallenbad*, [1968]
Nr. 18: *Bewährtes Material im Stil der Zeit*, [1969]
Nr. 19: *Resopal in Schulen und Sakralbauten*, [1969]
Nr. 20: *Der Laden – die Visitenkarte der Konsumgesellschaft*, [1970]
Thema: Krankenhäuser, [1972]
Nr. 23: *Thema: Sportstätten*, [1974]

Resopal GmbH (Hg.): *125 Jahre Forbo-Resopal. Design ist unser Werkstoff. Das Buch zum Jubiläum*, Groß-Umstadt 1992

Resopal GmbH (Hg.): Resopal® 2008, Groß-Umstadt 2003
Resopal GmbH (Hg.): Resopal® 2004 – Casual Wear, Groß-Umstadt 2004
Resopal GmbH (Hg.): Resopal® 2005 – Qwear, Groß-Umstadt 2005
Resopal GmbH (Hg.): Resopal® 2006 – Soft Wear, Groß-Umstadt 2006

Schmiedeke, Carl W./Müller, Maik/Hiller, Mathias: *Züge der Berliner S-Bahn. Die eleganten Rundköpfe. Baureihe 477, Bauarten Bankier, Olympia, 1937-41 und Peenemünde*, Berlin 2003

Schriefers, Thomas/Ernst, Ekkehard (Hg.): *Jupp Ernst. Werk und Lehre, 70 Jahre Designgeschichte*, Hagen 2000

Stadt Darmstadt (Hg.): *Helmut Lortz – Mittendrin und nebenbei. Mit Texten von Georg Hensel und Bernd Krimmel*, Darmstadt 1990

www.deutsches-kunststoff-museum.de

www.resopal.de

www.sintetica.de

■ **ABBILDUNGSNACHWEIS**
PHOTO CREDITS

Bauhaus-Archiv Berlin/Fotostudio Bartsch: S. pp. 32-33; 34 o. t.
Bergische Universität Wuppertal, Fachbereich F, Kunst- und Designgeschichte: S. pp. 25, 3; 57 M. c. + u. b.
Bildagentur Hamburg/Michael Fisler-Wohlert: S. pp. 150 l.
Brandenburgisches Landeshauptarchiv, Potsdam: S. pp. 34 u. b. (2)
Braun GmbH: S. p. 63, 4
Deutsches Technikmuseum Berlin: S. pp. 28-29; 111, 4
Deutsches Historisches Berlin: S. p. 64
Kathrin Ehmer, Berlin: S. pp. 83 u. b.; 124, 2
Freie Universität Berlin: S. p. 90
Galerie in der Kinderklinik, Universitätsklinikum Freiburg: S. p. 136
Dr. Jupp Gauchel, Karlsruhe: S. pp. 138-139
Germanisches Nationalmuseum, Nürnberg, mit freundlicher Genehmigung Courtesy: Anna Renate Biermann-Ernst: S. pp. 22-23; 30; 35 l. (3); 37 o. t.; 42-43; 45; 46 l.; 47; 52; 54-55; 56/57; 57 o. t.; 58-59; 72, 1
Gerhard Hagen, Bamberg: S. p. 83 r. (3)
Michael Heinrich, München: S. pp. 154; 160-161
Oliver Heissner, Hamburg: S. pp. 162-163
Helle, Jochen/artur: S. p. 150 r.; 153 r.
Hempel, Joerg/artur: S. p. 149 r.
Christoph Hess Fotodesign, Frankfurt am Main: S. pp. 77, 2; 134/135 o. t.
Historische S-Bahn e. V., Berlin: S. pp. 37 u. b. + r.; 66, 3 (4)
Werner Huthmacher, Berlin: S. pp. 130-131
Institut für Stadtgeschichte, Frankfurt am Main: S. pp. 66, 1 + 2; 86, 1
Andreas Kelm, Darmstadt: S. pp. 134 u. b.
Jo Klatt, Hamburg: S. pp. 113 l. u. b.; 121 u. b.
Bettina Knittel, www.knittelbild.de: S. pp. 96 r.; 118 o. t. l.
Luuk Kramer, Amsterdam: S. pp. 164-165
Kunststoff-Museums-Verein e.V., Düsseldorf: S. pp. 20; 26 M. c.; 27 M. c. + r.;
Dr. Dr. h. c. Günter Lattermann: S. pp. 13; 14; 15; 17; 18; 19
Rudolf Lübben, Mönchengladbach: S. pp. 77, 3; 80 (5); 81; 118/119 u. b.; 120 o. t. (2); 121 o. t. (3); 122 u. b.; 122/123 o. t.; 128-129;
Stefan Müller, Berlin: S. pp. 88-89; 90 (2); 90/91
Markus Palzer, Darmstadt: S. pp. 82 (3); 82-83; 166-167
Firma Poggenpohl: S. p. 111, 5
Monika Pötter, Wuppertal: 50-51
Dr. Ingrid Skiebe, Düsseldorf: S. p. 142
Stadtarchiv Montabaur: S. p. 74, 1
Staubach, Barbara/artur: S. p. 149 l.
Sigurd Steinprinz, Wuppertal: S. pp. 140-141
Tom Stark, Frankfurt am Main: S. pp. 143; 178
Stiftung Deutsches Rundfunkarchiv, Wiesbaden: S. p. 24, 1
ullstein bild, Berlin: S. pp. 25, 4; 26 l.; 120 u. b.
Vereinigte Spezialmöbelfabriken VS, Tauberbischofsheim: S. p. 85
von Haussen, Christoph/artur: S. p. 148 l.
Peter Whamond/Collector›s Guild, www.germanmilitaria.com: S. p. 25 r.
Wilsonart International, Temple, Texas (USA): S. pp. 126-127; 170

Abbildungen aus Publikationen Illustrations from publications
Berliner Philharmonisches Orchester (Hg.): Philharmonischer Almanach, Bd. 2, Berlin 1983: S. p. 151 l.
Braun-Feldweg, Wilhelm: Normen und Formen industrieller Produktion, Ravensburg 1954: S. p. 49
Constanze – Schöner wohnen – Schöner bauen, Nr. 27, IV 1956: S. pp. 104-105
Herzog & de Meuron 2002-2006. In: El croquis, 129/130, Barcelona 2006: S. p. 151 r.
Jüdisches Museum Berlin (Hg.): Architekt Daniel Libeskind. Dresden, 1999: S. p. 152 (2)
Krewinkel, Heinz W.: Kaum wiederzuerkennen. Das Haus der Glasindustrie in Düsseldorf. In: Deutsche Bauzeitung, 125 (1991), Nr. 7: S. p. 152 M. c.
Landesgewerbemuseum Stuttgart (Hg.): Industrie und Handwerk schaffen. Neues Hausgerät in USA. Erste Ausstellung neuzeitlicher Gebrauchsgegenstände aus USA in Stuttgart vom 20.3.-25.4.1951: S. p. 41 l.
Morgan, Conway Lloyd: Show me the future. Engineering and design by Werner Sobek, Ludwigsburg 2004: S. p. 152 r.
Müller-Wulckow, Walter: Architektur 1900-1929 in Deutschland. Reprint und Materialien zur Entstehung, Königstein im Taunus 1999: S. p. 148 r.
Nerdinger, Winfried: Walter Gropius, Berlin 1985: S. p. 147 (3)
Rapaport, Brooke Kamin/Stayton, Kevin L.: Vital Forms. American Art and Design in the Atomic Age 1940-1960, New York 2001: S. p. 40 (2)
Sayah, Amber: Neanderthal Museum, Mettmann. In: Domus, Nr. 789, (1997): S. p. 153 l.
Schneider, Romana/Nerdinger, Winfried/Wang, Wilfried: Architektur im 20. Jahrhundert. Deutschland, München 2000: S. p. 146 r.
Schwabe, Amtor: Was fängt der Architekt mit den Kunststoffen an? In: baukunst und werkform, Heft 5, 1953: S. pp. 41 r.; 44 (2)

Sämtliche hier nicht aufgeführten Abbildungen stammen aus dem Archiv der Resopal GmbH, Groß-Umstadt.
All illustrations not mentioned individually belong to the Resopal company archives, Gross-Umstadt.

■ AUTOREN

Dr. Eva Brachert
geboren 1959, Kulturanthropologin, von 1987 bis 1991 und seit 2004 als Restauratorin für Gemälde und Skulpturen am Landesmuseum Mainz tätig.

Dr. Gerda Breuer
geboren 1948, Kunsthistorikerin, seit 1996 Professorin für Kunst- und Designgeschichte an der Bergischen Universität Wuppertal.

Peter Cachola Schmal
geboren 1960, seit 1991 freischaffender Architekt, Architekturpublizist mit Lehraufträgen an der TU Darmstadt und FH Frankfurt, ab 2000 Kurator beim DAM, seit April 2006 Direktor des DAM.

Dr. Ingeborg Flagge
Studium der Archäologie und Ägyptologie in Köln, Rom und London, Chefredakteurin von „Der Architekt", Bundesgeschäftsführerin des Bundes Deutscher Architekten (BDA), Professorin für Baugeschichte an der HTWK in Leipzig, von 2000 bis 2005 Direktorin des DAM.

Ulrich Höhns
geboren 1954, freier Architekturhistoriker und -kritiker, seit 1992 wissenschaftlicher Leiter des Schleswig-Holsteinischen Archivs für Architektur und Ingenieurbaukunst (AAI).

Dr. Dr. h.c. Günter Lattermann
geboren 1943, Diplomchemiker, seit 1978 Akademischer Direktor am Fachbereich Makromolekulare Chemie I an der Universität Bayreuth, seit 1990 Sammlung alter Kunststoffe, gründete 2005 die Deutsche Gesellschaft für Kunststoffgeschichte, deren Vorsitz er innehat.

Gerd Ohlhauser
geboren 1948, Industrie-Designer, von 1975 bis 1990 freiberuflich und als Dozent für technisches Design tätig, von 1991 bis 1997 Designmanager bei Perstorp Decorative Laminate, seit 1998 verantwortlich für das Design der Resopal GmbH.

Romana Schneider
geboren 1952, Architekturhistorikerin, Mitarbeiterin bei der Bauausstellung Berlin 1984, Redakteurin bei „Architectural Design" (London) und »Domus« (Mailand), von 1990 bis 2000 Kuratorin beim DAM, seit 2000 freiberufliche Kuratorin.

Gert Selle
geboren 1933, emeritierter Professor für Theorie, Didaktik und Praxis der ästhetischen Erziehung an der Universität Oldenburg, lebt als Kulturhistoriker und Essayist in München. Seit 1973 hat er regelmäßig Bücher und Aufsätze über Kunstpädagogik, Design und Wohnen publiziert.

■ AUTHORS

Dr. Eva Brachert
Born in 1959. Cultural anthropologist; worked as a picture and sculpture restorer at the Landesmuseum in Mainz from 1987 to 1991 and again since 2004.

Dr. Gerda Breuer
Born in 1948. Art historian; since 1996 professor of art and design history at the Bergische University, Wuppertal.

Peter Cachola Schmal
Born in 1960. Since 1991 architect in private practice; architectural writer lecturing at the TU Darmstadt and FH Frankfurt; from 2000 curator at DAM; since April 2006 director of DAM.

Dr. Ingeborg Flagge
Studied Archaeology and Egyptology in Cologne, Rome and London; editor-in-chief of "Der Architekt;" chief executive at the Bund Deutscher Architekten (BDA); professor of architectural history at the HTWK in Leipzig; from 2000 to 2005 director of DAM.

Ulrich Höhns
Born in 1954. Freelance architectural historian and critic; since 1992 academic director of the Schleswig-Holstein Archiv für Architektur und Ingenieurbaukunst (AAI).

Dr. Günter Lattermann
Born in 1943. Chemist; since 1978 academic director at the Department of Macromolecular Chemistry I at the University of Bayreuth; in 1990 began a collection of historical plastics; in 2005 he founded the Deutsche Gesellschaft für Kunststoffgeschichte (German Society for the history of plastic), which he currently chairs.

Gerd Ohlhauser
Born in 1948. Industrial designer; from 1975 to 1990 worked both freelance and as a lecturer in technical design; from 1991 to 1997 design manager with Perstorp Decorative Laminate; since 1998 responsible for design with Resopal GmbH.

Romana Schneider
Born in 1952. Architectural historian; worked for IBA Berlin 1984; on the editorial team of "Architectural Design" (London) and "Domus" (Milan); from 1990 to 2000 curator at DAM; since 2000 freelance curator.

Gert Selle
Born in 1933. Emeritus professor for the theory, didactics and practice of aesthetic education at the University of Oldenburg; now lives and works in Munich as a cultural historian and essayist. Since 1973 he has regularly published books and articles on art education, design and domestic interiors.

A

Aalto, Alvar   120
Adler, Friedrich   19
AEG   110
Agrolinz Melamine International   183
Albamit   13
Albert (Firma)   13
Albertit   13
Albertol   13
Alboresin   12, 13, 16, 28
Alfred Nobel & Co.   16, 34
Alresat   13
Aminoplast   11, 12, 14, 16, 17, 18, 28, 31, 34, 171
APK GmbH   74/75
Architekten SZZ   82/83
Arjo Wiggins   183
Arp, Jean   40
Artschwager, Richard   133
Atelier Gebel   82/83, 124
Auböck, Carl   114

B

Baekeland, Leo Hendrik   11, 183
Bakelit   11, 14, 21, 171, 172, 173, 183
Bakelit Hares   21
Bakelite AG   183
Balzer, Dieter   7
BAR Architekten   179
Bartning, Otto   39, 148
Baudrillard, Jean   177
Bauschdekor   183
Bertoia, Harry   120
Brandlhuber, Arno   152, 153
Braun AG   61, 62/63, 101, 113, 120/121
Braun, Max   99, 100
Braun-Feldweg, Wilhelm   48, 49
Brinckmann-Schmolling, Ute   134/135
Brown, Boveri & Cie. AG Mannheim (BBC)   15, 16, 18, 113
Bumb & König (Beka)   12

C

Carl Zeiss Jena   25
Carola Schäfers Architekten   88/89, 90/91
Casein »Galalith«   41
Cellophan   41
Celluloid   11, 39, 41, 48, 177
Claus en Kann Architecten   164/165
Contzen, Lars   110

D

Dell, Christian   12, 13, 14, 15, 16, 17, 18, 19, 32/33, 172
DEMAG   78
Deutsche Werkstätten   53, 99
Diolen   13
Dresser, Christopher   42, 43, 44, 45
Duropal   110
Duroplast   180
Dynamit AG   12, 16, 34, 96

E

Eames, Charles   40
Eames, Ray   40
Ebonit   41
Egger   110
Eiermann, Egon   53, 150
Ernst, Jupp   15, 17, 18, 22/23, 24, 30, 36, 42, 45, 46, 49, 50, 53, 54/55, 56/57, 58/59, 61, 72, 98, 103, 109, 133, 137

F

Fagus-Werke   147
Formica   95, 100
Foucault, Michel   51

G

G. M. Pfaff Nähmaschinen   99, 100
Gauchel, Jupp   138/139
Getalit   100
Glatfelder-Konzern   183
Gretsch, Hermann   19
Grieshaber, HAP   86/87, 133, 136/137
Gropius, Walter   13, 18, 147
Gruco   95, 110

H

H. Römmler AG   11, 12, 14, 15, 16, 17, 18, 21, 23, 24, 25, 26, 27, 28, 29, 31, 32, 34, 35, 53, 96, 98, 173
H. Römmler GmbH   11, 17, 18, 34, 44, 45, 46, 47, 48, 53, 54, 56, 62, 65, 94, 98, 99, 100, 101, 103, 106, 123, 133, 137
H2S Architekten   82/83, 166/167
Hares   11, 13
Harnstoff   16
Harnstoffharz   11, 14, 18, 65
Hartmann, Dr. Gustav B. von   46
Helming, Helene   27
Herzog, Jacques & de Meuron, Pierre   151
Heubach, Friedrich   174
Hexion Speciality Chemicals   183
Hild und K Architekten   154, 160/161
Hirzel, Stephan   46
Höger, Fritz   149
Hutton, Louisa   152, 153

I

IG Farben   16, 149
Ingres, Dominique   183
Interprint   183
Jeanneret, Pierre   147, 148

J

Junghans   113

K

Kalderoni   129
Kallmorgen, Werner   151
Karbolit-Werke   17
Kersting, Walter Maria   13, 19
Knoll International   99, 100, 120
Kombinat Duroplast-Halbzeuge   17
Krauss, Julius   152, 153
Krimmel, Bernd   86/87, 134/135
Kunstharz   11, 12, 19, 21, 27, 133, 172, 180, 183
Kunststoff   7, 11, 12, 14, 15, 18, 19, 21, 34, 35, 39, 40, 41, 46, 47, 48, 49, 50, 51, 54, 93, 94, 95, 96, 97, 98, 99, 100, 101, 103, 160, 171, 172

L

Lachaise, Gaston   40
Lasertype   7
Le Corbusier   147, 148
Leowald, Georg   53
Libeskind, Daniel   151, 152
Linoleum   177, 178
Löffelhardt, Heinz   46
Loire, Gabriel   150
Loos, Adolf   146, 183
Lortz, Helmut   76, 106/107, 122/123, 137
Lübben, Rudolf   76/77, 80/81, 118, 120, 123, 128/129

M

Mallet-Stevens, Robert   149
Margarete Steiff GmbH   146, 147
Masa   183
MeadWestvaco   183
Meladur   17
Melaminharz   17, 18, 65, 183
Meyer, Adolf   147
Mies van der Rohe, Ludwig   147, 148
Möbel und Textilien (Firma)   99, 100
Moltopren   13
Moore, Henry   40
Morris, William   42, 44, 146
Müller, Gerd Alfred   113

N

Netzwerkarchitekten 130/131
Neudecker, Irene 7
Neue Gemeinschaft der Wohnkultur e.V. 97
Niemann Architekten 162/163
Nitroseide 41

O

Oberhoff, Ernst 49, 50, 133
Oberhoff, Hans 74/75
Olms, Hans 131
Olms, Romana 131

P

Pelikan, Sarah 140/141
Pfau, Bernhard 151, 152
Phenol-Formaldehyd-Harz 11
Phenolharz 154, 183
Phenol-Kresol-Harz 31
Phenoplast 12, 13, 14, 17, 28, 31, 172
Phenoplast-Lackharze 13
Plastica GmbH 14, 32
Plastik 46, 49, 50, 51
Plexiglas 48, 49, 142
Poelzig, Hans 149
Poggenpohl 110
Pollack, Fritz 16
Pollopas 16, 34, 95
Polyester 40
Porzellanfabrik Rosenthal 99, 100
PVC 50, 51

R

Radio- und Elektrogeräte / Frankfurt 99, 100
Rasch Tapeten 100
Resofloor 13
Resoform 13
Resopal GmbH 7, 21, 65, 127, 183
Resoplan 13, 158, 166
Rheinisch-Westfälische Sprengstoff AG 16
Römmler, Arthur 11
Römmler, August Hermann 11
Römmler, Hermann jun. 11, 12
Römmler, Kurt 11
ROTAR 142
Rudolf Lang (Firma) 25
Ruskin, John 42, 44, 145, 146
Rüttgers, Julius 183

S

Saarinen, Eero 40
Sauerbruch, Matthias 152, 153
Schaefer, Donald 7
Scharoun, Hans 53, 151
Schattdecor 183

Schellack 11
Schmittel, Wolfgang 60/61, 118/119
Schoeller und Hoesch 183
Schricker, Rudolf 178
Schwabe, Amtor 41
Schwippert, Hans 8, 27, 40, 47, 50, 53
Sedlmayr, Hans 41
Seeger, Mia 46
Selb 99, 100
Semper, Gottfried 41, 42, 145, 146
Sils & Co. 25
Sisal-Teppiche 99, 100
Sobek, Werner 152
Sprelacart 17, 66/67, 171
Stark, Tom 143, 177, 178
Steiff, Richard 147
Stickstoffwerke Ostmark AG 183
Stotz-Kontakt 15, 16, 18, 19
Styropor 13
Süddekor 183
Sullivan, Louis 45

T

Taut, Bruno 151, 152
Tauwerk-Fabrik 99, 100
TOYO kitchen & living 110

U

Ultrapas 34, 95, 100
Ungers, Oswald Mathias 149, 150

V

VEB Plasta 17
VEB Pressstoffwerke Spremberg 17
Venditor Kunststoff Gesellschaft 34, 96
Vulcanit 11
Vulkanfiber 41

W

Waca 97, 113
Wagenfeld, Wilhelm 18, 19
Wermann, Armin 87
Wilson Plastics 127
Wilson, Ralph Senior 127
Wilson, Sunny 127
Wilsonart International 127
WK-Möbel 53
Württembergische Metallwarenfabrik 99, 100

Z

Zamp Kelp, Günter 152, 153
Zelluloid 49
Zellulose 133
Zellulose-Azetat 41

■ **IMPRESSUM**
   IMPRINT

Dieses Katalogbuch erscheint anlässlich der Ausstellung
»Original Resopal. Die Ästhetik der Oberfläche«
vom 25. November 2006 bis 11. Februar 2007
veranstaltet vom Deutschen Architekturmuseum
Dezernat Kultur und Wissenschaft
Stadt Frankfurt am Main
This catalogue is published to accompany the exhibition
"Original Resopal. The Aesthetics of Surface"
November 25, 2006 until February 11, 2007
organized by the Deutsches Architekturmuseum
Department of Culture and Science
Frankfurt/Main, Germany

Konzeption und Redaktion Conception and editing
Romana Schneider
Kurztexte Short texts
Romana Schneider
Übersetzung ins Englische Translation from the German
Chris Charlesworth
Lektorat Copy editing Katharina Döring, Rachel Hill
Fotos Sammlungsgegenstände Photos of collection items
Ingmar Kurth, Frankfurt am Main
Scans und Bildbearbeitung Scans and digital imaging
Lasertype GmbH, Darmstadt
Ausstellungsgestaltung Exhibition design
DESERVE GbR, Raum und Medien Design, Wiesbaden/Berlin
Öffentlichkeitsarbeit Public relations
Ursula Kleefisch-Jobst
Sekretariat, Verwaltung, Leihverkehr Administrative staff, registrar
Inka Plechaty, Jeanette Bolz, Anke Gabriel
Aufbau und Hängung Setting up and hanging
Verantwortlich Responsible: Paolo Brunino, Enrico Hirsekorn, Detlef Wag-
ner-Walter. Leitung Under the direction of Christian Walter
Leihgeber Lenders
Prof. Dr. Gerda Breuer, Bergische Universität Wuppertal; Bauhaus-Archiv
Berlin; Margarete d'Hooghe, Darmstadt; Druckerei Ph. Reinheimer, Darm-
stadt ; Galerie in der Kinderklinik, Zentrum für Jugendmedizin und Kin-
derheilkunde am Universitätsklinikum Freiburg; Galerie Netuschil, Darm-
stadt; Dr. Jupp Gauchel, Karlsruhe; Historische S-Bahn e. V., Berlin; Egbert
Kaiser, Köln; Prof. Helmut Lortz, Darmstadt; Prof. Dr. Thomas Müller, VS
GmbH & Co, Tauberbischofsheim; Gerd Ohlhauser, Darmstadt; Sarah
Pelikan, Wuppertal; Monika Pötter, Wuppertal; Resopal GmbH, Groß-
Umstadt; Ronald Seffrin, Direktor Schuldorf Bergstraße, Seeheim-Jugen-
heim; Dr. Ingrid Skiebe, Düsseldorf; Tom Stark, Frankfurt am Main
Fotografen Photographers
Fotostudio Bartsch, Berlin; Kathrin Ehmer, Berlin; Gerhard Hagen, Bam-
berg; Michael Heinrich, München; Oliver Heissner, Hamburg; Christoph
Hess, Frankfurt am Main; Werner Huthmacher, Berlin; Andreas Kelm,
Darmstadt; Bettina Knittel, www.knittelbild.de; Luuk Kramer, Amster-
dam; Ingmar Kurth, Frankfurt am Main; Peer Lübben, Bonn; Stefan Müller,
Berlin; Markus Palzer, Darmstadt; Steinprinz fotodesign, Wuppertal
Architekten Architects
Claus en Kaan Architecten, Amsterdam; Atelier Gebel, Berlin; Hild und K
Architekten, München; H2S Architekten, Darmstadt; Netzwerkarchitek-
ten, Darmstadt; Niemann Architekten, Hamburg; Carola Schäfers Archi-
tekten, Berlin; Architekten SZZ, München

Danksagung für großzügige finanzielle Unterstützung
Acknowledgements for generous funding
Resopal GmbH, Groß-Umstadt
Lasertype GmbH, Darmstadt
Für spezielle Hilfe danken wir Acknowledgements
Anna Renate Biermann-Ernst, Kreuzau-Untermaubach; Dr. Ulrike Enke,
Institut für Geschichte der Medizin, Gießen; Dr. Regina Fiebich, Stadt-
archiv Montabaur; Germanisches Nationalmuseum, Nürnberg; Sabine
Hartmann, Bauhaus-Archiv Berlin; Peter Hartwein, Braun GmbH, Kron-
berg; Mathias Hiller, Berlin; Theo Hofsäss, Freiburg; Horst Kaupp, Braun
GmbH, Kronberg; Jo Klatt, Hamburg; Bernd Krimmel, Darmstadt; Dr. Hel-
mut Laun, Seeheim-Jugenheim; Rudolf Lübben, Mönchengladbach; Maik
Müller, Berlin; Christoph Rau, Photographie + Filmlocation Germany,
Darmstadt; Prof. Wolfgang Schmittel, Kronberg; Uta Scholten, Kunststoff-
Museums-Verein e.V., Düsseldorf; Erich Wagner, Deutsches Architektur-
museum, Frankfurt am Main; Dr. Klaus Weber, Bauhaus-Archiv Berlin

Gestaltung und Satz Design and setting Sven Schrape
Umschlaggestaltung Cover design Sven Schrape, Gerd Ohlhauser
Frontispiz Frontispiece Dieter Balzer: Installation DAM, 2006
Hintergründe Backgrounds Macassar, Matt Alu Dots, Ivory, Entwurf von
Jupp Ernst Design by Jupp Ernst, Dekor der sechziger Jahre 1960s design,
Coal Concrete, Chill Terrage
Lithografie Lithography bildpunkt, Berlin
Druck und Bindung Printing and binding GCC Grafisches Centrum Cuno,
Calbe

Die Deutsche Bibliothek verzeichnet diese Publikation in der Deutschen
Nationalbibliografie; detaillierte bibliografische Daten sind im Internet
über http://dnb.ddb.de abrufbar. Die Deutsche Bibliothek lists this publication in the Deutsche Nationalbibliografie; detailed bibliographic data
are available in the Internet at http://dnb.ddb.de

jovis Verlag
Kurfürstenstraße 15/16
10785 Berlin

www.jovis.de

ISBN 3-939633-04-6 (Buchhandelsausgabe)
978-3-939633-04-4